The Lightweight Treated Soil Method

The Lightweight Treated Soil Method

New Geomaterials for Soft Ground Engineering in Coastal Areas

Takashi Tsuchida

Department of Social and Environmental Engineering,
Hiroshima University,
Hiroshima, Japan

and

Kazuhiko Egashira

Coastal Development Institute of Technology,
Tokyo, Japan

CRC Press
Taylor & Francis Group
Boca Raton London New York

CRC Press is an imprint of the
Taylor & Francis Group, an **informa** business
A BALKEMA BOOK

CRC Press
Taylor & Francis Group
6000 Broken Sound Parkway NW, Suite 300
Boca Raton, FL 33487-2742

First issued in hardback 2019

ISBN-13: 978-90-5809-692-0 (hbk)

Visit the Taylor & Francis Web site at
http://www.taylorandfrancis.com

and the CRC Press Web site at
http://www.crcpress.com

Library of Congress Cataloging-in-Publication Data

A Catalogue record for this book is available from the Library of Congress

Contents

Foreword

Every year, seaport and offshore airport construction projects use large quantities of soil excavated from mountainous areas near the sites. However, large-scale excavation for soil supply is becoming extremely difficult for environmental reasons. Meanwhile, the lack of sites for disposing of large quantities of dredged soil and surplus construction soil which are generated annually is becoming a serious problem in most coastal cities. Accordingly, a new geo-material, which is made of dredged soil or waste soil and which has sufficient engineering properties and is available at a reasonable cost, has been strongly demanded by engineers involved in construction projects in coastal areas.

With this background, a research group organized by the Soil Mechanics Laboratory of the Port and Harbour Research Institute developed the lightweight treated soil (LWTS) method. In this method, dredged soil or waste soil with a high water content is mixed with lightweight material and is chemically treated to enable it to be used as a backfilling or reclamation material in onshore construction works. The wet density of the lightweight soil is usually adjusted to $1.0\,g/cm^3$ above sea level and $1.1–1.2\,g/cm^3$ below sea level. The conventional design shear strength ranges from $200\,kN/m^2$ to $400\,kN/m^2$. The LWTS method in coastal soft ground areas can directly reduce the consolidation settlement and earth pressure of gravity structures and also helps to reduce the cost and time for ground improvement and foundation works. As the chemically treated soil is usually more expensive than natural soils taken from mountains, its light weight is a useful property for making the treated soil more economically viable.

The first large-scale application of the lightweight treated soil method was for part of the restoration work following the Great Hanshin-Awaji Earthquake Disaster of 1995. Since then, a total of $277{,}000\,m^3$ of lightweight treated soils has successfully been used in 23 seaport and airport projects in Japan as of 2002. Commercially, the LWTS method is called the SGM Lightweight Soil Method, where SGM is the abbreviation of "Super Geo-Materials".

In 1999, the Coastal Development Institute of Technology in Japan has edited and published a technical manual for the lightweight treated soil method for Japanese engineers. This book is the English version of that manual, and also includes recent case studies of construction work in which the LWTS method was used. This book was written by researchers at the Port and Airport Institute and research engineers who are the members of the technical committee of the SGM lightweight Soil Method Association. The book describes the properties, design and construction techniques of LWTS and various applications, based on extensive research and experience in Japan.

Finally, we wish to thank all authors and colleagues who assisted the editing work. We sincerely hope that this book will serve as a useful reference for all those involved in soft ground engineering around the world.

Takashi Tsuchida
Professor, Department of Social and Environmental Engineering, Hiroshima University
Former Head of Soil Mechanics Laboratory,
Port and Harbour Research Institute

Kazuhiko Egashira
Doctor of Engineering, Executive Director,
Coastal Development Institute of Technology

List of Authors

Takaaki Yoshida	TOYO Construction Co., Ltd.
Kazuhiro Yamamura	TOA Corporation
Hiroshi Shinsha	PENTA-OCEAN Construction Co., Ltd.
Masahiko Kuwabara	FUDO Construction Co., Ltd.
Hideki Sakanoi	WAKACHIKU Construction Co., Ltd.
Hidetaka Kudo	KUBOTA Construction Co., Ltd.

Chapter 1

Outline of the SGM Lightweight Soil Method

1.1 INTRODUCTION

In the construction projects of seaports, large amounts of mountain soils, river and sea sands have been used for the reclamation or backfilling of quaywalls. However, it is becoming increasingly difficult to obtain these soils especially due to the recent issues of environmental protection. On the other hand, a large volume of soft clayey soils is dredged annually and dumped in waste disposal sites enclosed by seawalls in Japanese seaports. Figure 1.1 shows the change of waste volume dumped during the period of 1980–1995. As shown in the figure, approximately 18 million m^3 of the wastes were disposed of every year, of which 45% was dredged soil, 21% was surplus soils from construction sites in land and 8% was industrial waste. Recently, the lack of waste disposal sites in seaports has become a serious problem and about 30 billion yen has been expended annually so far on the construction of seawall structures for new disposal sites. Moreover there have been strong calls due to social and environmental aspects for the reuse of dredged soils in port and harbor construction works.

Most of the dredged soil is clayey soils with high water content and is too weak to be used for fill-materials without any treatment. However, chemically treated dredged soil is

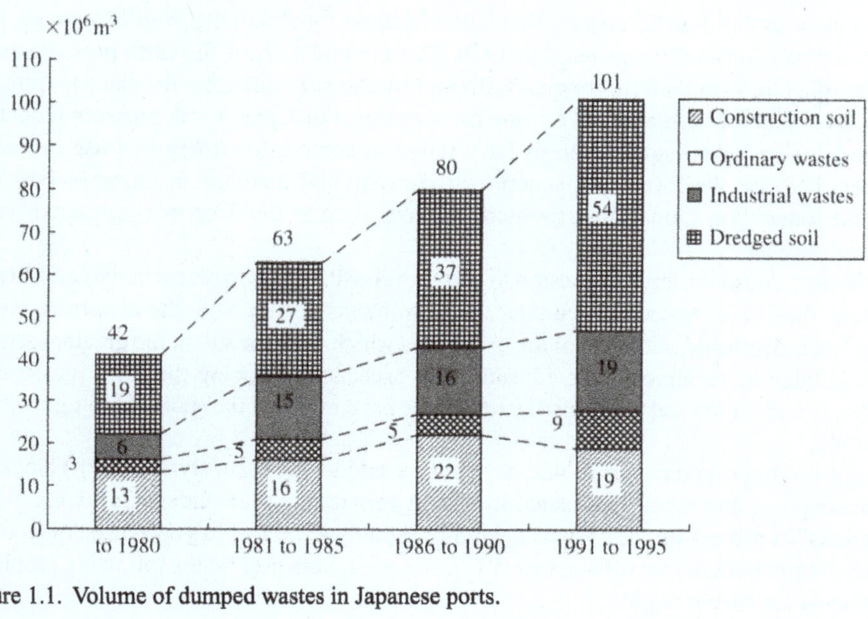

Figure 1.1. Volume of dumped wastes in Japanese ports.

usually more expensive than natural soils taken from the hills or mountains because of the cost for machinery and equipment. In order to make treated soil more competitive with natural soils, some additional advantages are necessary. The light weight is an effective property to make treated soil more economically viable.

In recent projects in coastal areas, as construction sites have become farther from the original shore lines, more difficult and complicated ground conditions at deeper sea water level have been encountered. In these conditions, geotechnical problems have arisen, such as the long-term settlement and lateral deformation of soft ground, higher susceptibility to earthquake disasters and the increased cost for ground improvement. The use of lightweight treated soil is an effective countermeasure to solve these problems.

In the lightweight treated soil (LWTS) method, the dredged clay of high water content is mixed with lightweight materials and chemically treated to be used for the backfilling of quaywalls or for reclamation or embankments in port and harbor works. The mixed lightweight materials are air foam or expanded polystyrol (EPS) beads with small diameters. The wet density of the lightweight soil is usually 1.0 g/cm^3 in the ambient condition and 1.1–1.2 g/cm^3 below sea level. The LWTS method was first applied for the rebuilding of a quaywall damaged by the Great Hanshin-Awaji Earthquake in 1995. Since then, the method has been used successfully in several seaports and airports projects in Japan. Commercially, the LWTS is called *SGM Lightweight Soil Method*, where SGM is the abbreviation of "Super Geo-Materials".

1.2 DEVELOPMENT OF THE LIGHTWEIGHT TREATED SOIL METHOD

Research on the use of lightweight materials for harbor construction dates back to the 1960s. As a result of this work, in 1971, the Port and Harbor Research Institute applied for a patent on the Earth Pressure Reduction Method for Retaining Wall Structures, and this patent was officially registered in 1976. This method reduces the earth pressure using lightweight blocks as backfill for quaywalls and similar structures, but the patent expired in 1991, as it had not been adopted for any harbor project in Japan. Earth pressure reduction methods using lightweight materials have failed to come into widespread use in harbor projects because they were not superior to conventional methods in terms of cost and because foundation ground improvement methods such as the deep mixing method have become popular.

However, due to the large expansion of offshore constructions in recent harbor and airport projects, they have started to encounter many problems dealing with the extreme increase of the water depth and thickness of the clay layers which would result in the greater increase of consolidation settlement. The advantageous aspects of reducing the earth pressure on structures and improving their seismic resistance have returned the spotlight to lightweight materials.

Lightweight ground materials that have been used include foam styrene (EPS) blocks in the atmosphere and water granulated iron-blast-furnace slag for underwater work. Water granulated iron-blast-furnace slag is the lightweight material that has been used most often as lightweight material for offshore work [1]. However, both present the following problems when used for harbor works.

Foam styrene blocks

- Their density is between 0.012 and 0.03 g/cm^3 providing them with superior lightness. However, this makes them float on the surface of water, preventing their offshore use.
- This method is more expensive than other methods.

Water granulated iron-blast-furnace slag

- The density ranges from 0.8 to 1.3 g/cm^3, however its underwater unit weight γ' is 7.0 kN/m^3. Therefore, not much weight reduction can be expected.
- Because it is a by-product from the steel making process, the amount of its supply is confined to the production volume.
- There might be limits on the supply of material when there is a sudden demand for a large quantity.

Since about 1990, the disposal of construction waste soil and dredged soil has become a serious social problem.

In response to these circumstances, Port and Harbor Research Institute, Coastal Development Institute of Technology, and 23 private companies set up the Research Committee on the "Development of Lightweight Mixed Ground Material for Harbor and Marine Environments (SGM Lightweight Soil Research Committee)" in 1992 (then chaired by Okumura Tatsuro, Professor of Okayama University). This committee has been actively involved in the development of new lightweight ground material with priority on the following five points:

- Appropriate lightness
- Environmental safety
- Reuse of construction waste soil and dredged soil
- Underwater placing
- Large volume of supply with economical advantage within a short period time

During this development process, the first large-scale application ever in Japan using the lightweight treated soil method was made as a part of the restoration work following the Awaji Hanshin Earthquake of 1995.

1.3 CHARACTERISTICS AND TYPES OF LIGHTWEIGHT TREATED SOIL METHODS

A lightweight treated soil is a ground material with a density between 0.6 and 1.5 g/cm^3 made by adding water to dredged soil or to other source soils mixed with lightener material (bubbles, beads), stabilizing agent, etc. The mixture is very fluid while it is being mixed or placed. However, by the reaction with the stabilizing agent the material hardens with strength properties equal to or superior to good quality soil materials.

The following are the characteristics of the lightweight treated soil method:

- Lightweight treated soil is, unlike with a natural ground material, a homogenous ground material whose density and strength can be adjusted as necessary.
- It is a construction method that has little effect on the nearby marine environment, because its mix proportion can be designed to provide underwater separation resistance.
- Efficient reuse of dredged soil and construction waste soil with high water content is possible.

- Maintaining the specified quality, even under water, can be ensured by the underwater separation resistance of the mixture.
- Due to its flowability, it can be placed in any shape desired without compaction using only pump feed pressure. Therefore, a large volume of the material can be placed in a short period of time.

The lightweight treated soil is classified into air foam treated soil and beads treated soil according to the type of lightener (air bubbles, beads) that is used. The following are the characteristics of these methods.

Characteristics of air foam treated soil

Its density is adjusted by varying the quantity of bubble material mixed with the soil. The density shows an increasing tendency by defoaming of the air bubbles before hardening and by the water pressure during underwater curing.

The more stabilizing agent that is added to the mixture, the greater its unconfined compressive strength (q_u). It has strength properties very similar to those of stiff clays. If water pressure or other pressure acts on the soil during curing, the density increases and the rising density is accompanied by a slight increase in q_u.

Its permeability is almost identical to the coefficient of permeability for dredged soils that have been hardened by a stabilizing agent, and the quantity of lightener has little effect on it.

Regarding the effects of the water content of the adjusted slurry, the higher its water content, the greater its flowability.

Also, within a normal range, the lower its water content, the greater its underwater separation resistance. Its separation resistance is also influenced by various factors like soil quality, type of lightener, quantity of stabilizing agent added, type of foaming agent, type of dilution water, and the placing method.

Characteristics of beads treated soil

Its density is adjusted by varying the quantity of EPS beads added to the soil. Within a normal range, its density upon mixing and after underwater curing shows a consistent value.

As in the case of the air foam treated soil, the more stabilizing agent that is added to the soil, the greater its unconfined compressive strength (q_u), and its strength properties are almost identical to those of stiff clay. However, in a normal range, the value of q_u is almost unaffected even if water pressure and other pressures act on the soil during curing.

Its permeability and flowability are almost identical to those of air foam treated soil.

Its underwater separation resistance is, as in the case of the air foam treated soil, influenced by the water content of the adjusted slurry, the soil quality, the type of lightener, the quantity of stabilizing agent added, and the placing method. It is highly influenced by increases in the grain diameter of the beads.

1.4 USES OF THE CONSTRUCTION METHOD

The lightweight treated soil method can be used as a material for backfilling, land reclamation, embankments, refilling and so forth.

Use as backfilling material

Because of its favorable characteristics of earth pressure reduction, it can be used to reduce the sections of gravity quaywalls and of sheet piles. This means it is efficient in reducing the construction period and ultimately results in overall cost savings. Specifically, it is applied to the cases shown in Figure 1.2.

- *Gravity quaywall.* It can reduce the size of the caisson section because it can reduce the earth pressure acting on the back of the caisson. This can reduce the size of the foundation mound and narrow the range of the soil improvement.
- *Sheet pile quaywall.* Because it can lower earth pressure acting on the back of sheet piles, it can reduce the section of the sheet piles and shorten their embedding depth. This can reduce the size of the section of the counterfort retaining pile.

Use as land reclamation material

Due to the lower effective weight of SGM lightweight soil, consolidation settlement can be reduced, therefore it can be used efficiently as a reclamation material on top of soft ground. Specifically, it can be used in the cases shown in Figure 1.3.

- *Raising quaywalls.* Because it can lighten the reclamation soil, it can minimize the settlement of reclaimed ground. It can also reduce the earth pressure applied to the raised part.
- *Forward extension of existing quaywall.* It can lighten the reclamation soil to reduce the earth pressure acting on a new bulkhead. When the bottom layer is formed with soft soils, it can minimize the consolidation settlement caused by the reclamation load.

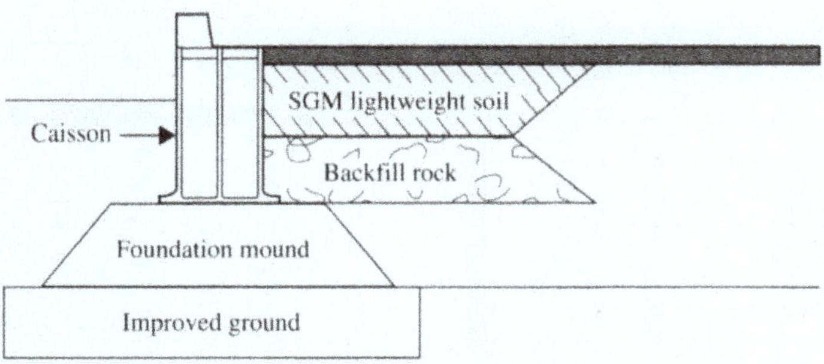

Figure 1.2a. Example of the case for backfilling material.

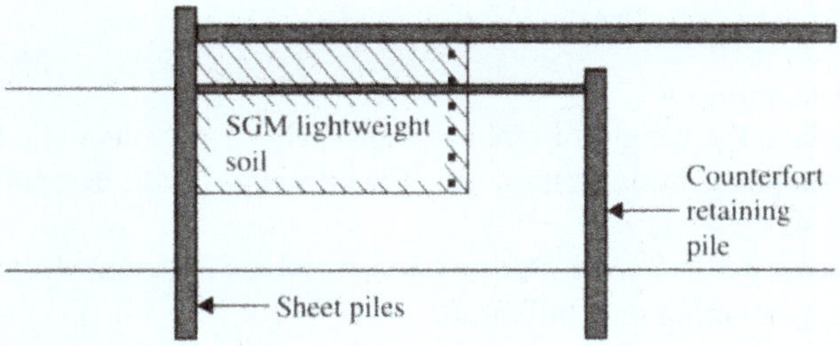

Figure 1.2b. Example of the case for backfilling material.

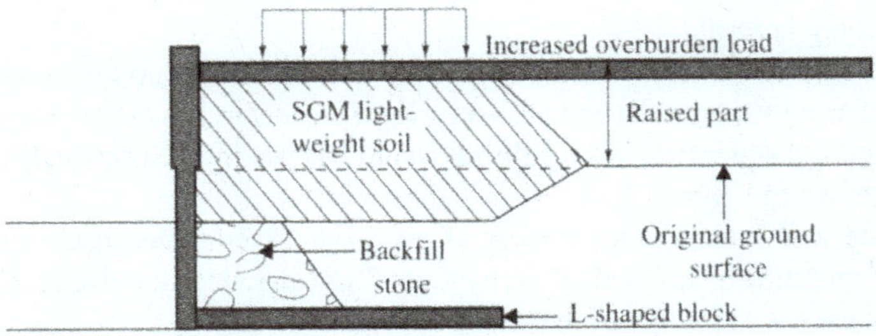

Figure 1.3a. Example of the case for reclamation soil.

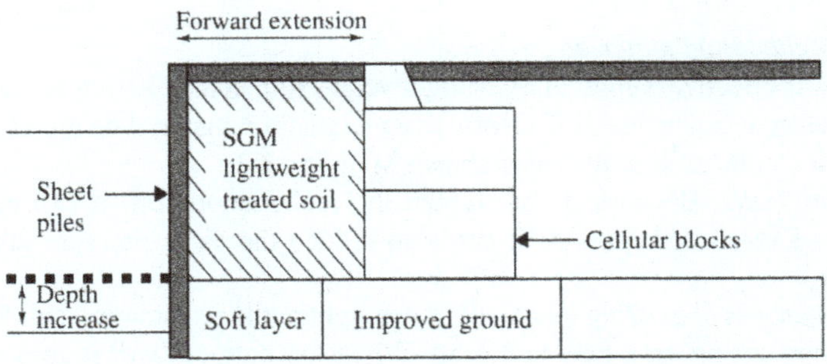

Figure 1.3b. Example of the case for reclamation soil.

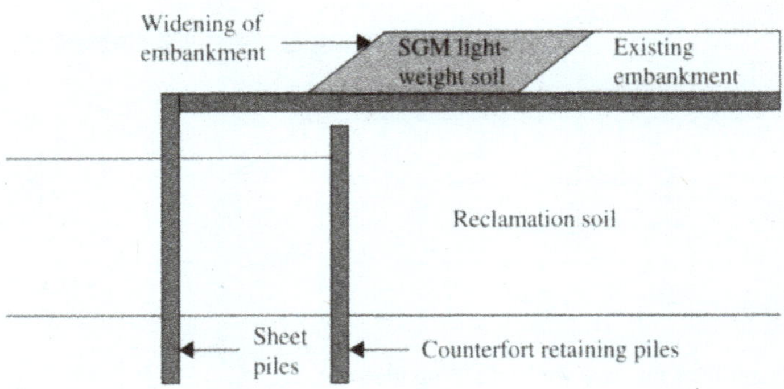

Figure 1.4a. Example of the case for embankment material.

Use as embankment material

Because it can reduce the weight of embankment soil, it can reduce the load on existing embankments, structures, buried pipes, etc. Specifically, it can be applied to the cases shown in Figure 1.4.

- *Embankment behind a bulkhead.* Because it can lighten embankment soil, it can lower the overburden load acting on a bulkhead.
- *Embankment on soft ground.* Because it can lighten embankment soil, it can restrict settlement, and because SGM has a certain degree of strength, it prevents sliding failure.

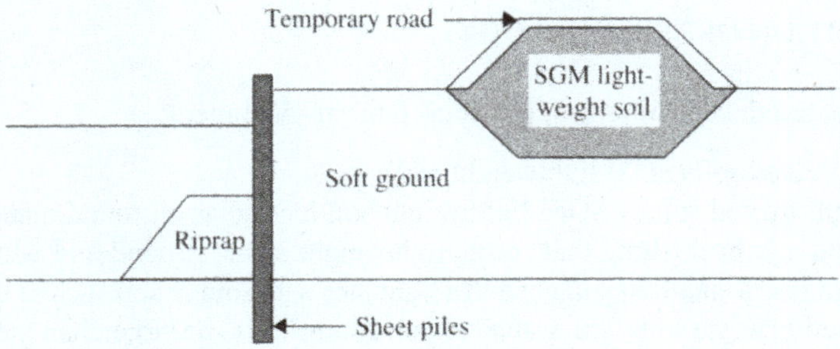

Figure 1.4b. Example of the case for embankment material.

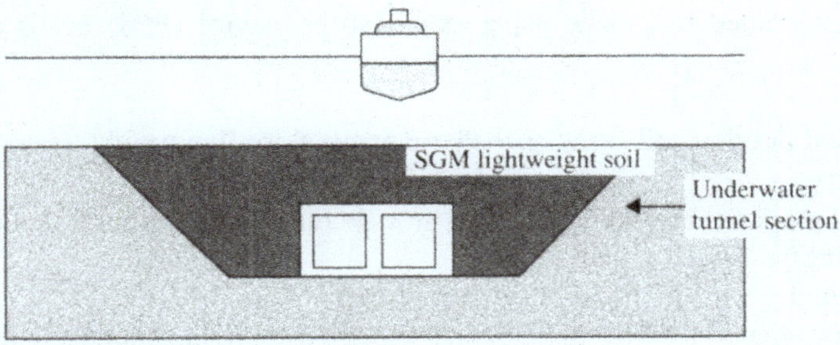

Figure 1.5a. Example of use as refilling material.

Figure 1.5b. Example of use as refilling material.

Use for refilling

Because it resists separation underwater, it can be used as an underwater refilling material. Also, due to its flowability, it can be used for efficient refilling work in parts with small intervals between structures. Specifically, it can be used for the cases shown in Figure 1.5.

- *Refilling over immersed tunnels*. It can be used for underwater construction because of its underwater separation resistance. It can reduce the load acting on the immersed tunnel sections. Due to its flowability, efficient refilling work is possible.
- *Refilling of existing structures*. Using SGM lightweight treated soil for refilling to build a new structure on top of the existing structure, can reduce the load acting on the existing structure.

1.5 DEFINITION OF TERMINOLOGY

The following are the definitions of terms used in this document:

Lightweight treated soil (SGM lightweight soil)
: Lightweight treated soil or SGM lightweight soil is used as a ground material for land reclamation and backfilling that forms lightweight stable ground and which has been made by mixing a stabilizing agent and a lightener with source soil such as dredged soils or construction waste soils with water content controlled to be larger than the liquid limit.

Air foam treated soil
: Lightweight treated soil with air foam mixed as its lightener.

Beads treated soil
: Lightweight treated soil made using expanded polystyrol (EPS) beads mixed as its lightener.

Source soil
: Dredged soil or other soils used as the base material for lightweight treated soil.

Adjusted slurry
: Slurry soil prepared by adding water to provide the source soil with specified density and flowability.

Stabilizing agent
: Cementitious material added to guarantee specified strength. Another stabilizing agent can be used as necessary.

Lightener
: Material mixed to lower the weight of ground material, and in this report, refers to either air foam or to polystyrol beads.

Air foam
: Material made with a foaming material and air and mixed with the material to lighten it. In principle, the prefoam method is used.

Foaming material
: Material made by diluting a foaming agent with water (sea water, fresh water) prior to foaming.

Foaming agent
: Material that is the basic ingredient of foaming material. Surface active agents and animal protein type are the principal kinds in use.

Defoaming rate
: During the mixing to prepare the air foam treated soil, the foam partly disappears as it is broken, shrunk, and separated during mixing, placing, and other processes during and after construction. This is called defoaming, and the volume of defoamed foam/volume of mixed foam ratio is called the defoaming rate.

Expanded polystyrol beads
: Material that is mixed in the same way as air foam to lighten the material. It refers to lightweight grains made by foaming artificial resins. In this document, it refers specifically to EPS beads.

Additives (agents)
: Material that is used to improve execution properties such as material separation and pumpability of the lightweight treated soil. It includes bentonite, separation prevention

agents, flowability agents, water reducing agents, and coagulation promotion agents and retarders.

Treated soil

In this document, it is used as an abbreviation of lightweight treated soil.

Treated ground

In this document, it is used as an abbreviation of lightweight treated ground.

Super Geo Material (SGM)

A general term for new ground material for harbor and airport work that provides added values such as light weight (or heavy weight), safety (non-polluting), and recycling.

Slice method

A method of assuming a slip surface behind a structure and dividing the block of ground between the slip surface and the structure wall into slices in order to calculate the earth pressure based on the equilibrium of the weight, buoyancy, shear strength of the slip surface, and forces produced by the seismic force of each slice.

Failure mode

To perform earth pressure calculations, three modes are assumed for a case where a slip surface passes through treated soil. Mode I is a linear slip passing through the interior of the treated soil, mode II is slippage accounting for hypothetical shearing in the treated soil, and mode III is slippage that occurs along the boundary with the treated soil.

Flow value

An index (mm) that represents the flowability of treated soil, the flow value in the atmosphere is based on the cylinder method stipulated by JHS A 313 and the flow value under water is obtained by placing a specimen on the bottom of a water tank and pouring 16 cm of artificial sea water into the tank, then removing the cylinder to measure its flowability.

Surface active agent

This is a general term for materials that are dissolved in a fluid in order to sharply increase the surface tension of the fluid. Like oil and water, they do not mix, and can solubilize many fluids.

1.6 SYMBOLS

A : foaming agent mass
α : length of the arm of the slip circle center of external force (H)
B : minimum width of the foundation
B_{min} : minimum width of the lightweight treated soil
BP : back pressure equal to the curing pressure
B : width of the slice
c : cohesiveness
c_i : cohesiveness of layer i
c_u : undrained shear strength
c_0 : undrained shear strength of clay on the foundation surface
d_i : depth
D : embedding depth of the foundation
E_{50} : modulus of deformation
E_i : lateral force acting on the right side of the slice

F : safety factor
G : shearing modulus
G_0 : initial shearing modulus
H : *lateral external force acting on the block of soil in the slip circle*
h : damping constant
h : thickness of the layer
h : thickness of the layer where the earth pressure is sought below the residual water level
h_i : thickness of layer i
h_i : thickness of the layer of soil above the residual water level
h_j : thickness of the layer of soil higher than the layer where the earth pressure is sought below the residual water level
I_p : plasticity index
K_a : active earth pressure coefficient
K_{ai} : active earth pressure coefficient of layer i
K_0 : earth pressure coefficient at rest
k : seismic intensity
k' : apparent seismic intensity
k_h : lateral seismic intensity
L : length of the foundation
l_i : length of the slip surface of the slice
l : length of the bottom of the slice
m_s : mass of the dried soil
m_w : (sea) water mass
m_c : stabilizing agent mass
m_A : foaming agent mass
m : foaming multiplier
N : loading frequency
N_{co} : bearing strength coefficient
N_i : vertical strength of the slip surface
n : dilution multiplier
n : form factor of the foundation
P_a : active earth pressure strength
P_{ai} : active earth pressure strength acting on the wall at the bottom surface of layer i
P_y : consolidation yield stress
q_a : allowable bearing strength
q_u : unconfined compressive strength
q_{max} : maximum axis differential stress during compression
R : radius of the slip circle
T_i : shear strength of the slip surface
V_i : vertical force acting on the right side of the slice
V_0 : volume of air at atmospheric pressure
V_h : volume of the bubbles at depth h
V_a : air volume in lightweight treated soil
V_{a0} : quantity of air in the foam
V_b : volume of the expanded polystyrol beads
W : total mass of the slice

W_i : total mass of the slice
W'_i : effective mass of the slice
w : unit load per unit volume of the ground surface
w_1 : liquid limit
x : lateral distance between the center of gravity of the slice and center of the slip circle
α : increment rate
α : slip surface angle
β : angle formed by the ground surface and horizontal
γ : shear strain
γ : unit weight of the soil
γ_t : wet unit weight of the soil
γ_i : unit weight of layer i
γ' : underwater unit weight of the soil
γ_2 : unit weight of the soil of the ground above the bottom surface of the foundation
δ : angle of friction of the wall surface
ΔV_{a0h} : volume compressibility of air included in foam at water depth h
ΔV_{a0} : equivalent air volume at atmospheric pressure
ΔV_a : foam increase
ΔV_h : volume compressibility of air at water depth h
ε : compression strain
ε_a : axial strain
ε_v : volume strain
ε_h : shrinkage strain in the axial direction
ε_r : shrinkage strain at right angles to the axis
ζ_i : angle formed by the failure surface and the horizon at layer i
ζ_a : angle formed by the failure surface and the horizon
θ : composite angle at earthquake $\theta = \tan^{-1} k$ or $\theta = \tan^{-1} k'$
μ : coefficient of friction
ν : Poisson's ratio
ρ : density
ρ_s : density of the soil grains
ρ_w : (sea) water density
ρ_c : density of the stabilizing agent
ρ_t : density of the adjusted slurry
σ : compressive stress
σ_a : axial stress
σ'_c : confining pressure
ϕ : internal friction angle
ϕ : composite angle at earthquake $\phi = \tan^{-1} k$ or $\phi = \tan^{-1} k'$

REFERENCES

1. Ministry of Construction, Technical Survey Section of the Minister's Secretariat and the Public Works Research Institute: Outline of Methods Developed to Utilize Construction Soil, Public Works Research Center, March 1995.

Chapter 2

Engineering Properties of Lightweight Treated Soil

2.1 OUTLINE

Lightweight treated soil is a ground material with a density between 0.6 and 1.5 g/cm^3 made by mixing sea water, lightener, stabilizing agent, etc. with dredged soil or other source soil. Lightweight treated soil has flowability immediately after it is mixed, but the reaction with the stabilizing agent ultimately produces hardened soil with strength properties equal to or superior to good quality soil material. This chapter describes the properties of the materials used for the mixture, and the characteristics before and after hardening, particularly density, flowability, resistance to separation underwater, permeability, and strength properties.

To design lightweight treated soil, it is necessary to fully clarify these basic properties of lightweight treated soil.

Figure 2.1 shows an outline of the materials used to make lightweight treated soil and the mixing procedure.

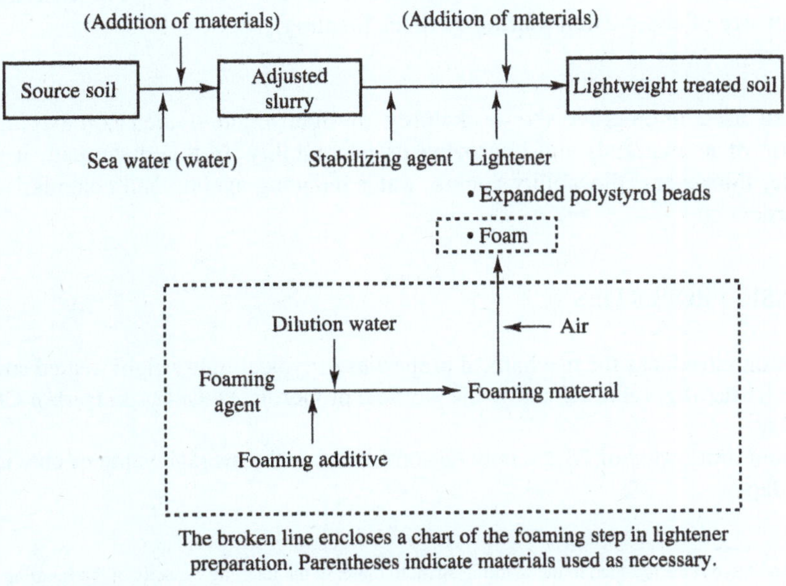

Figure 2.1. Outline of the materials used.

The terminology in this chapter is used as defined in the previous chapter. It is necessary to pay close attention to the following items concerning the lightener.

Air foam
Because defoaming is increased when the source soil has a high organic material content, great care is taken to add a foaming additive* to increase the quantity of foaming material used.

Foaming agent
Material used to manufacture the air foam. It is categorized as the surface active agent type and animal protein type. The main constituents of surface active agent type are alkyl–ethyl or anion compounds. The main constituents of the animal protein type are the products of the hydrolysis of keratin protein such as animal horns and hooves. Generally, the surface active agents provide superior flowability and pressure resistance. It has also been reported that both satisfy standards stipulated in laws and regulations concerning the content of harmful substances [1].

Both have good foaming properties even when sea water is used as the dilution water, and a foaming agent that limits the defoaming after preparation of the lightweight treated soil should be used.

Expanded polystyrol beads
A comprehensive technical term for extremely lightweight granular solid resin made by using a heat source such as steam to heat and foam synthetic resin. It includes foam styrene beads (EPS). This document deals with EPS beads.

To use this material under water, its specifications must be decided accounting for its buoyancy, deformation caused by water pressure, workability, etc.

Experiments have shown that it should have an apparent density of at least 0.02 g/cm^3 and grain size of about 2 mm (during 50 times foaming).

Additives
These are used to improve the workability of lightweight treated soil by preventing separation of its materials and improving its pumpability. In this document, it refers to bentonite, thickeners, flowability agents, water reducing agents, and coagulative agents and retarders.

2.2 BASIC PROPERTIES

This section introduces the mechanical properties of typical lightweight treated soil before and after hardening. Table 2.1 shows the physical properties of Kawasaki Harbor Clay from Tokyo Bay.

Its liquid limit value of 76.1% may be considered as the average value of clay in harbor areas in Japan.

* Foaming additive: This is added to the foaming agent to increase the foaming capacity of the foaming agent and to boost the safety of the air foam. It includes nonionic water soluble polymers, etc.

Table 2.1. Physical properties of the test soil.

Type of soil	Soil grain density	Particle size distribution (%)				Liquid limit (%)	Plastic limit (%)
		Gravel	Sand	Silt	Clay		
Clay	2.631	0.3	6.6	77.1	16.0	76.1	44.8

Table 2.2. Specifications of the foaming agent.

Lightweight material	Foaming multiplier	Density
Air foam	25×	40 g/*l*

Table 2.3. Specifications of the EPS beads.

Lightweight material	Foaming multiplier	Expanded polystyrol beads density (g/*l*)	Unit weight (g/*l*)
Expanded polystyrol beads (EPS beads)	30×	53.3	33.3
	50×	332.0	20.0
	70×	22.9	14.3

Note 1: The mix proportion design was performed using the density of expanded polystyrol beads.
Note 2: The void ratio of the unit weight (density including the voids between the expanded polystyrol beads) during each foaming was 0.37.

2.2.1 Test conditions

The following are four sets of mix proportion conditions:

- Mix proportion 1: Target strength 200 kN/m^2, target density 1.1 g/cm^3
- Mix proportion 2: Target strength 400 kN/m^2, target density 1.1 g/cm^3
- Mix proportion 3: Target strength 200 kN/m^2, target density 1.2 g/cm^3
- Mix proportion 4: Target strength 400 kN/m^2, target density 1.2 g/cm^3

Source soil

The sample soil was Kawasaki clay. Its physical properties are shown in Table 2.1.

Lightweight materials

The lightweight materials used were air foam and expanded polystyrol beads. The foaming agent was a surface active agent, and the expanded polystyrol beads were EPS beads. The specifications of each material are shown in Tables 2.2 and 2.3.

Stabilizer

The stabilizer was blast furnace cement B.

Sea water

Artificial sea water was used.

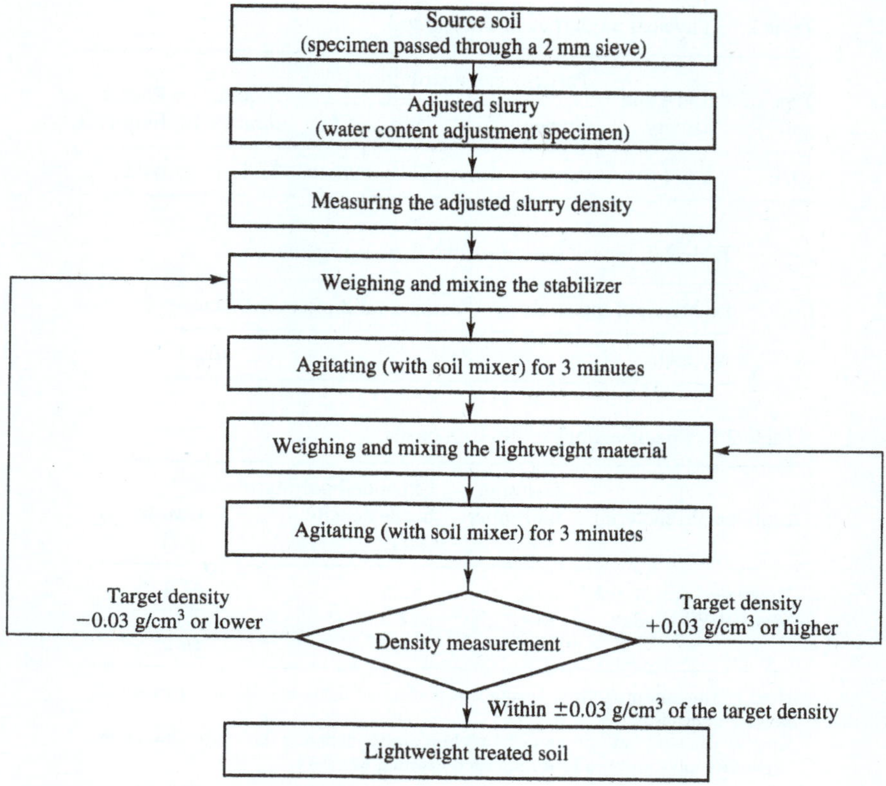

Figure 2.2. Preparation procedure for lightweight treated soil specimen.

Lightweight treated soil specimen preparation method
The specimens of lightweight treated soil were made by the procedure shown in Figure 2.2.

2.2.2 Properties of air foam treated soil

2.2.2.1 *Density properties*

Change in density in the atmospheric condition
The density of lightweight treated soil changes under the effects of the following factors:
- Defoaming before hardening
- Shrinkage caused by initial hardening
- Shrinkage caused by drying

Defoaming before hardening is the process by which the air foam defoams over time, and its extent depends on the type of source soils and its state after mixing.

The increase in density caused by volume shrinkage during initial hardening is about 0.03 g/cm³ according to previous observations [3]. The change in density caused by drying is a result of weight loss of water and solidification contraction, but if air foam treated soil poured in the atmosphere is covered with soil, drying can be neglected and its density

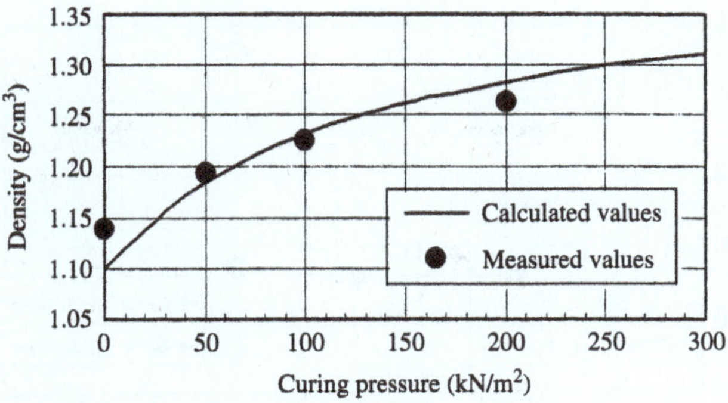

Figure 2.3. Curing pressure–specimen density relationship. [4]

Table 2.4. Mix proportion.

	Target density 1.1 g/cm^3		
Water content of adjusted slurry (%)	Adjusted slurry (kg/m^3)	Stabilizing agent (kg/m^3)	Air foam (l/m^3)
150	1,010	90	214.1

remains almost constant. The mix proportion design must consider in advance the density change caused by defoaming and initial hardening.

Change in density underwater

Because contraction of the air foam in air foam treated soil is caused by water pressure when the soil is placed underwater, the density becomes higher than when placing the soil in the atmosphere.

Figure 2.3 shows the curing pressure and specimen density relationship of air foam treated soil [4]. These experiment results show the density of the specimen assuming that contraction is not caused by initial hardening and that the volume of the foam inside the specimen decreases in accordance with Boyle's Law at a constant temperature under the effects of changes in curing pressure (but the internal pressure of the foam is assumed to be atmospheric pressure). This figure reveals that the density of the air foam treated soil is increased by the curing pressure until its value is almost identical to that calculated by Boyle's Law. Table 2.4 shows the air foam treated soil mix proportion used for this test.

Each test was carried out by using freshly prepared specimens placed in the mold inside a pressure vessel filled with sea water under the specified pressure for curing for 28 days at 20°C. Later the curing pressure was removed to form the specimen and then its density was measured.

Boyle's Law

At constant temperature conditions: $P_1V_1 = P_2V_2$

where, P_1, P_2: pressure; V_1, V_2: gas volume.

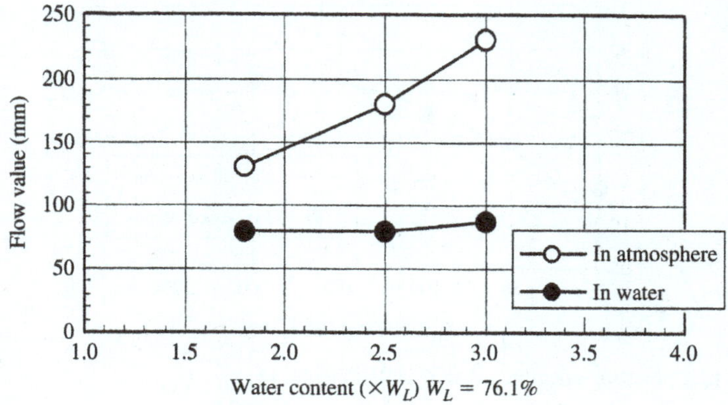

Figure 2.4. Water content–flow value relationship. [6]

Table 2.5. Mix proportion.

	Target density 1.1 g/cm^3		
Water content of adjusted slurry (%)	Adjusted slurry (kg/m^3)	Stabilizing agent (kg/m^3)	Air foam (l/m^3)
137 (1.8 W_L)	991	100	236
190 (2.5 W_L)	993	100	187
228 (3.0 W_L)	994	100	161

Density change underwater

It has been observed that if a specimen of air foam treated soil is cured for a long time in a water tank in a laboratory, water permeates gradually from the specimen surface, increasing the density. This occurs because deterioration advances from the boundary surface of the specimen and the water, and this deterioration is accompanied by a decrease in strength and an increase in density. The rate of deterioration is reported to be approximately 12.0 mm/year [18]. Therefore, it is necessary to cover the surface of air foam treated soil with soil, or to coat the surface with other material in order to avoid direct contact of the air foam treated soil with water. An investigation on the ground application of air foam treated soil revealed no evidence of an increase in density [5].

2.2.2.2 *Flow properties*

The flowability of air foam treated soil varies according to the water content and density of the adjusted slurry and the quantity of stabilizing agent.

Figure 2.4 shows the flow value of air foam treated soil in relation with the water content of adjusted slurry [6].

Table 2.5 shows the mix proportion of air foam treated soil in this experiment. The flow value in the atmosphere was measured in accordance with the JHS A 313 cylinder method, and the flow value underwater was measured by placing a specimen on the bottom of a water tank in the same way as in the atmosphere test, then pulling up a cylinder after filling the tank with artificial sea water to a depth of 16 cm.

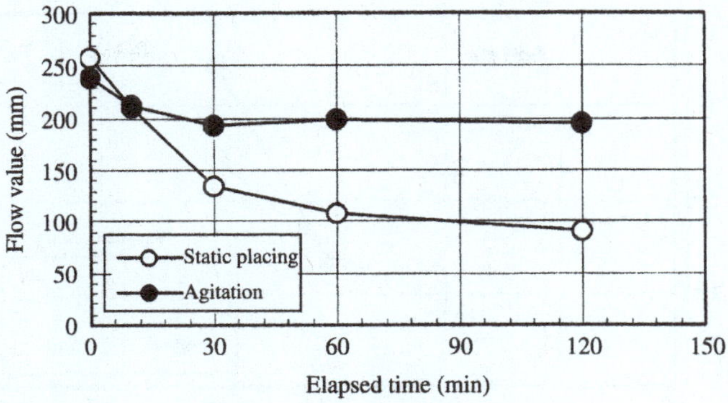

Figure 2.5. Elapsed time–flow value relationship.

As the figure shows, as the water content of the adjusted slurry increases, the higher the flow value. The flow value measured under water is lower and its variation caused by the water content is smaller than in the atmosphere because its self-weight is decreased due to the buoyancy. When the flow value in the atmosphere is 20 mm or less, it is self-sustainable in the water almost steady. The flowability is evaluated based on the flow value in the atmospheric condition.

Figure 2.5 shows the variation of flow values with the elapsed time from preparation of air foam treated soil.

In the static placing case, the test is carried out by placing the freshly prepared air foam treated soil in a static manner into the mixing-tray, and later taking some of the treated soil out from the mixing-tray at each specified measurement time to measure its flow value.

In the agitation case, the test is carried out by continuously agitating the freshly prepared air foam treated soil in a mixer and measuring the flow value at each measurement time.

As can be seen in the figure, in the static placing case, the flowability indicated by the flow value significantly decreases over time, while the flow value showed little decrease in the agitation case. Table 2.5 shows the mix proportions of the air foam treated soil used in this test.

2.2.2.3 *Underwater separation resistance*

Underwater separation resistance testing method

When lightweight treated soil is placed underwater, the material might be separated because it absorbs air and water and because of the viscous resistance of water itself, according to its mix proportion conditions, placing rate, and placing method. This separation could increase its density, lower its strength, and increase both turbidity and *pH*. The underwater separation resistance test [15][16][17] is a simple laboratory test that can evaluate the suitability of the mix proportion ratio by quantitatively clarifying the resistance to material separation of lightweight treated soil to be placed underwater.

The underwater separation resistance of air foam treated soil is evaluated based on the measured values of laboratory tests, but generally, the value of SS (Suspended Substances) is often used as the index. Figure 2.6 shows an example of the water content and SS of adjusted slurry. Table 2.5 indicates the mix proportion of air foam treated soil in this test. As this figure shows, the higher the water content of the adjusted slurry, the higher

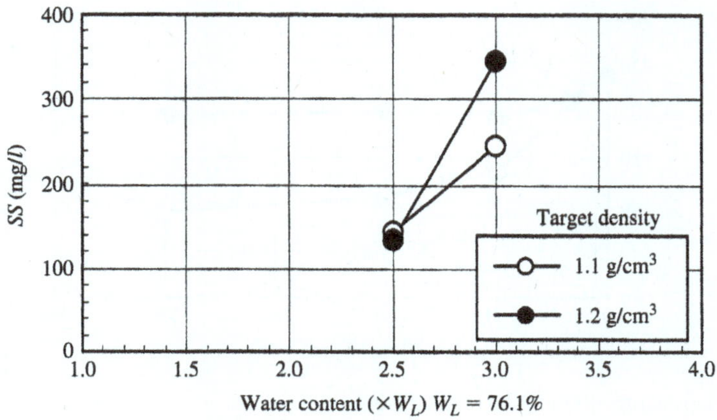

Figure 2.6. Water content–*SS* relationship.

the SS and lower the underwater separation resistance. This reveals that the underwater separation resistance is strongly influenced by the water content of the adjusted slurry. Separation resistance is also largely affected by soil quality, quantity of stabilizing agents and placing method as well. These issues remain to be solved through further researches and investigation.

Positioning of underwater separation resistance tests

Figure 2.7 shows the position of underwater separation resistance tests in the laboratory and an overall flow chart concluding with the determination of the basic mix proportion and placing rate based on the underwater separation resistance tests.

Test apparatus

Photograph 2.1 shows an outline of an underwater separation resistance test apparatus. It consists of a cylinder of perspex tube (internal diameter: 100 mm, height: 440 mm, internal volumetric capacity: 3,460 m^3), electric motor and piston, suction pump (nominal diameter: 1 inch), injection vessel (internal diameter: 200 mm, height: 200 mm, internal volumetric capacity: 6,280 m^3), and tip nozzle (internal diameter: 12 mm). It is constructed so that the rotation speed of the electric motor can regulate the placing rate (maximum placing rate is about 140 cm/sec).

Test procedure

The underwater separation resistance test is carried out by the following procedure:

- Artificial sea water of 300 m*l* is placed in the injection vessel.
- The outlet of the tip nozzle is anchored at a height of 12 mm (=1D) from the bottom surface of the injection vessel.
- The piston type cylinder is filled with the specimen of 3,000 m*l*.
- The cylinder is connected with the injection vessel through the pumping tube.
- The rotation speed of the electric motor is set to the specified pouring rate.
- The piston is operated to extrude the specimen and is then shut off when it is filled to the outlet of the tip nozzle.

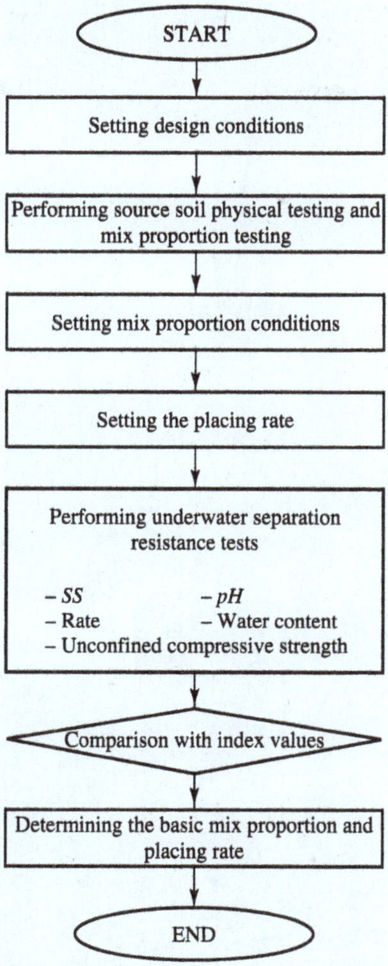

Figure 2.7. Positioning and overall flow of underwater separation resistance tests.

- The piston is operated again to continuously inject the specimen into the water in the injection vessel.
- The injection vessel is removed carefully, another injection vessel is prepared, and the procedure is repeated from step [1].
- About 1 minute after placing is completed, *SS* and *pH* values in the artificial sea water inside the injection vessel are measured.
- Each injection vessel is cured for 7 days in a standard curing condition which is performed either underwater or in the wet state at a humidity of almost 100% while maintaining the temperature at 20 ± 3°C.
- After curing, three specimens are sampled (if one specimen cannot be taken, at least two other specimens must be taken). These specimens are called underwater specimens.
- The density, water content, and unconfined compressive strength of the underwater specimens are measured.

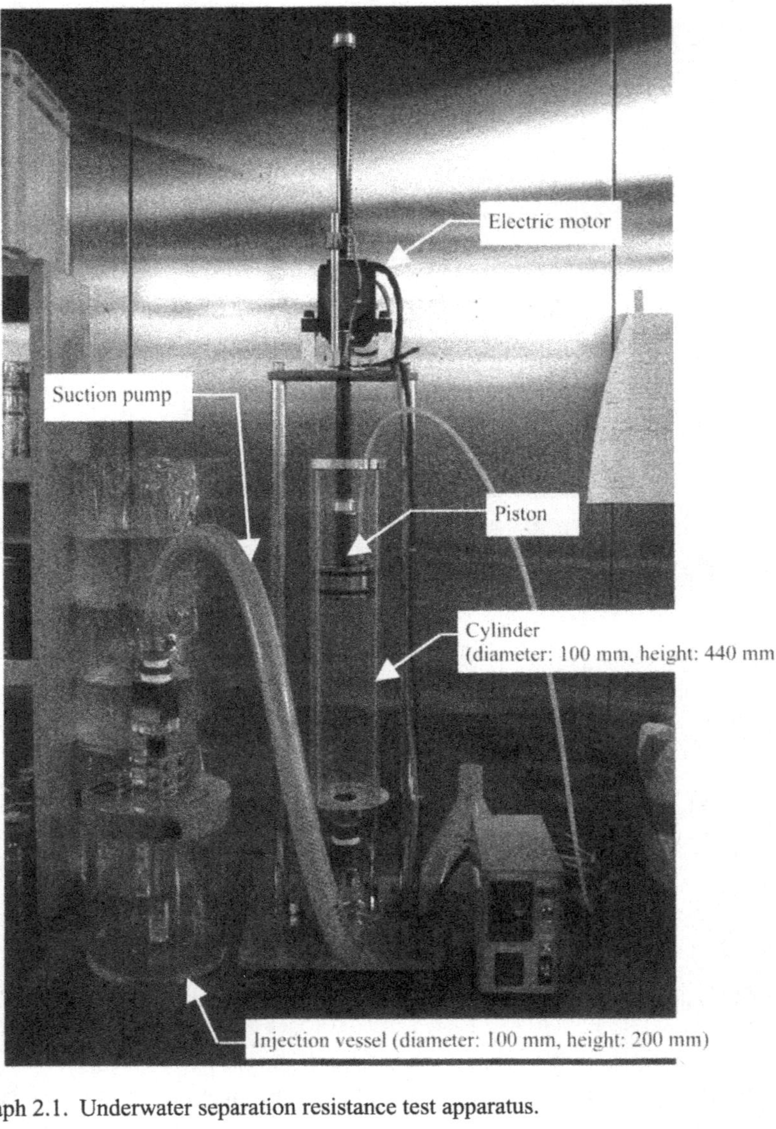

Photograph 2.1. Underwater separation resistance test apparatus.

Another three specimens are prepared with simple molds for each test, cured by standard curing for 7 days in the atmosphere, then their density, water content, and unconfined compressive strength are measured in the same way. Those tests should be performed under the conditions of an atmospheric flow value of 150 to 400 mm, and placing rate of 10 to 100 cm/sec.

Specimen preparation

Mix proportion conditions that can produce the specified material properties are selected based on the results of mix proportion tests and the specimen is prepared by the same method used for the mix proportion test. Additional numbers of specimens are to be prepared according to the number of test cases for different placing rate because it is necessary to prepare samples with a volume of $3\,l$ per test and three specimens for each underwater separation resistance test. The density and the flowability of the specimens prepared both change over time. The number of pouring rate cases is normally set at no more than five cases because about 5 minutes are required to perform each test. The density and atmospheric flow value are obtained by measuring these before and after the completion of the underwater separation resistance tests, organizing the results, and using the averages of these measured values.

Reporting test results
The test results are reported as the measured values of *SS* and *pH*, difference in the densities of the underwater and atmospheric specimens, the proportions of their water contents, and unconfined compressive strengths.

Reference indices
It is possible to set material property conditions that will prevent the lightweight treated soil from affecting the marine environment during field application, and constraints that will guarantee the specified material properties during underwater placing using the following indices for the underwater separation resistance test as criteria:

$$SS: SS < 100\,\text{mg}/l$$

$$pH: pH < 10.5$$

Underwater specimen density: atmospheric specimen density ±0.05 g/cm^3
Underwater specimen water content: increase rate of 10% or less against the atmospheric specimen water content *n*
Underwater specimen strength: 50% or more than the unconfined compressive strength of an atmospheric specimen of 7 days aging

The indices for *SS* and *pH* are for the condition that the marine area is closed during placing. Therefore, the conditions in the closed marine area should be reconsidered.

2.2.2.4 *Permeability*
Although the void ratio of air foam treated soil is extremely high, the effects of the existence of air foam on permeability are small, and this void ratio is almost the same as the ordinary cement treated soil without air foam. It has been reported that when 75 kg/m^3 of cement was added to source soil, the permeability decreased to between 1/100 and 1/10 of that in the case of no cement added [7].

2.2.2.5 *Strength properties after hardening*
Unconfined compressive strength
Figure 2.8 shows the variation of unconfined compressive strength (aged for 28 days) with different quantities of stabilizing agent.

Table 2.6 indicates the mix proportion of air foam treated soil during this test.

The figure shows that as the quantity of stabilizing agent increases, the unconfined compressive strength (q_u) linearly increases regardless of the water content of the adjusted

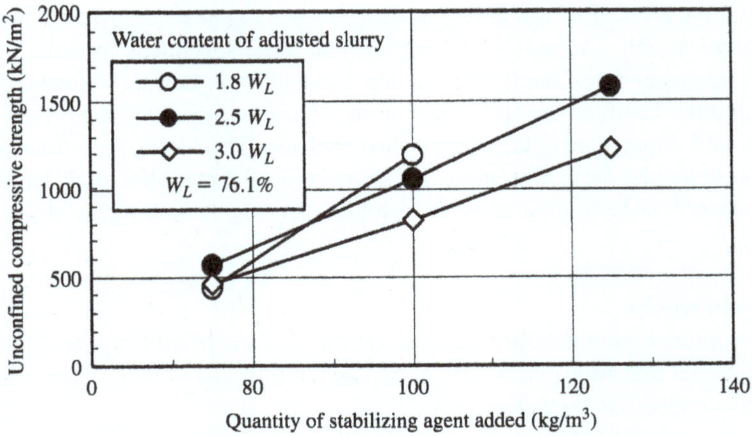

Figure 2.8. Quantity of stabilizing agent–unconfined compressive strength relationship.

Table 2.6. Mix proportions.

	Target density 1.1 g/cm³		
Water content of adjusted slurry (%)	Adjusted slurry (kg/m³)	Stabilizing agent (kg/m³)	Air foam (l/m³)
137 (1.8 W_L)	1,016	75	237
	990	100	247
	965	125	258
190 (2.5 W_L)	1,017	75	189
	992	100	200
	967	125	212
228 (3.0 W_L)	1,018	75	163
	993	100	175
	967	125	188

slurry. Furthermore, for the same quantity of stabilizing agent, q_u tends to show higher values as the water content of adjusted slurry decreases.

Figure 2.9 shows the variations of unconfined compressive strength with curing age. Table 2.7 shows the mix proportion of the air foam treated soil in this test. Strength development reaches more than 80% of the long-term strength (>90 days) at 28 days of aging.

Comparing the strength at 28 days of aging with that at 7 days, there is an increase of 1.5–2.0 times.

Figure 2.10 shows examples of stress–strain curves based on the unconfined compression test results.

The failure strain ranged from 0.5 to 1.5%. The mix proportion of air foam treated soil in this test is shown in Table 2.7.

Figure 2.11 shows the ratio of unconfined compressive strength of air foam treated soil by pressurized curing to that by atmospheric curing [8]. Table 2.7 shows the mix proportion of the air foam treated soil in this test. Pressurized curing increases the unconfined compressive

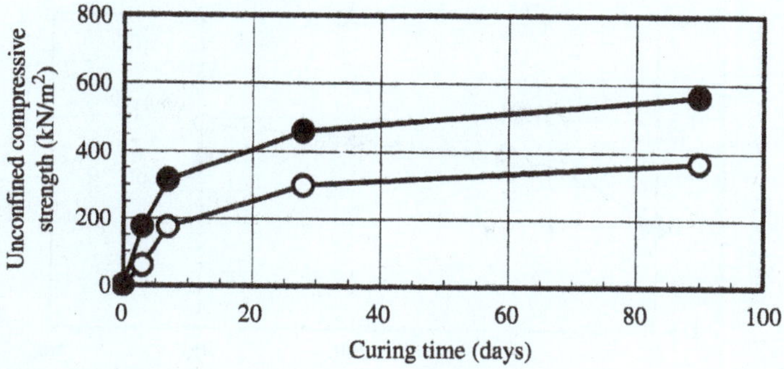

Figure 2.9. Curing time–strength relationship.

Table 2.7. Mix proportion.

Water content of adjusted slurry (%)	Target density 1.2 g/cm^3		
	Adjusted slurry (kg/m^3)	Stabilizing agent (kg/m^3)	Air foam (l/m^3)
190 (2.5 W_L)	1,028	67	91
	1,019	76	95

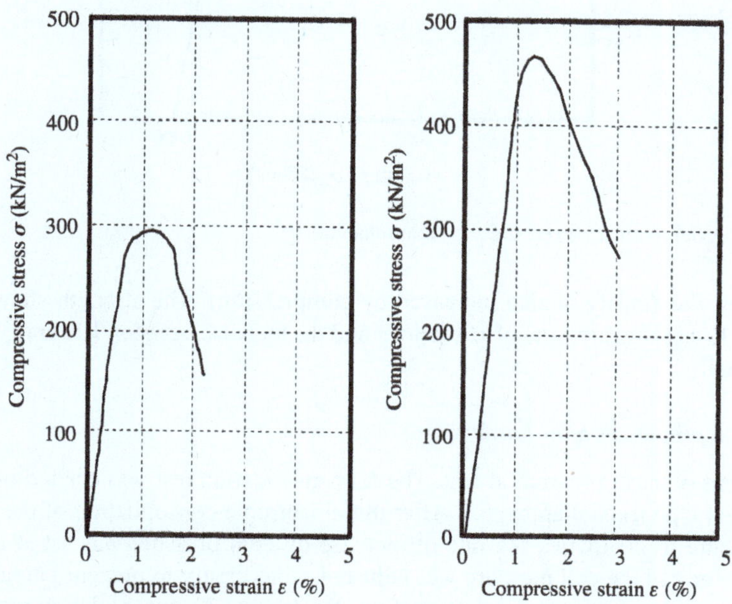

Figure 2.10. Stress–strain curves based on the unconfined compressive strength test.

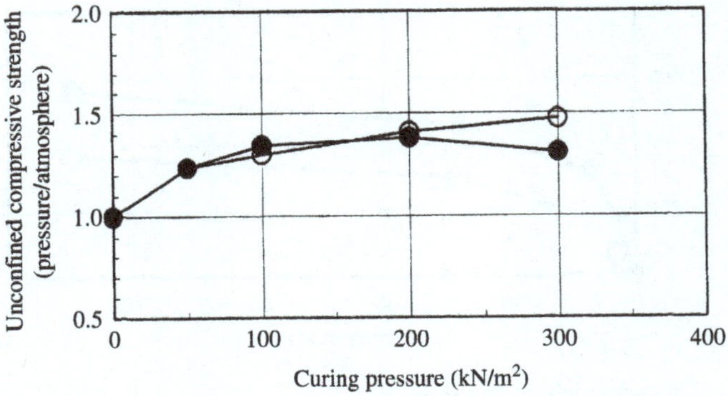

Figure 2.11. Curing pressure–unconfined compressive strength ratio relationship.

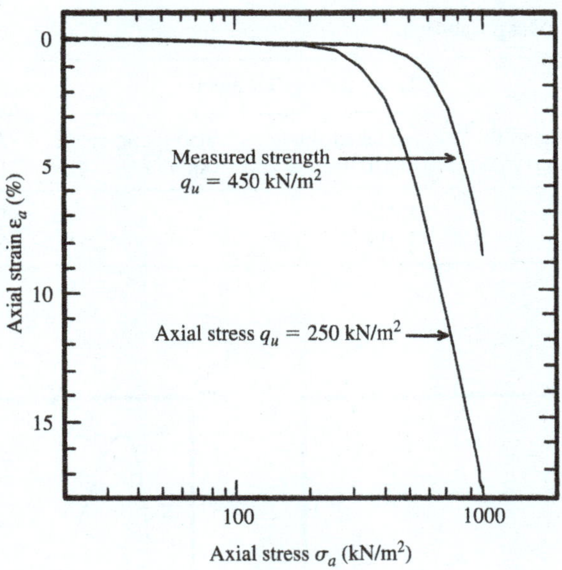

Figure 2.12. Axial strain ε_a–axial stress σ_a relationship.

strength (but the density is also increased by compression). The strength shows a large increase up to a curing pressure of 100 kN/m^2 and the increase between 100 and 300 kN/m^2 is rather small.

Compression properties after hardening

- Coefficient of earth pressure at rest. The K_0 consolidation test was carried out using a double cell type triaxial apparatus. After initial isotropic consolidation of the specimen at a pressure of 20 kN/m^2, the rate of increase of axial pressure was set at a constant 65 kN/m^2/hr and the cell pressure was adjusted as necessary to prevent lateral strain as the axial pressure was increased, to perform the triaxial K_0 consolidation test up to an axial pressure of 1,000 kN/m^2. Figure 2.12 summarizes the axial strain ε_a–axial stress σ_a

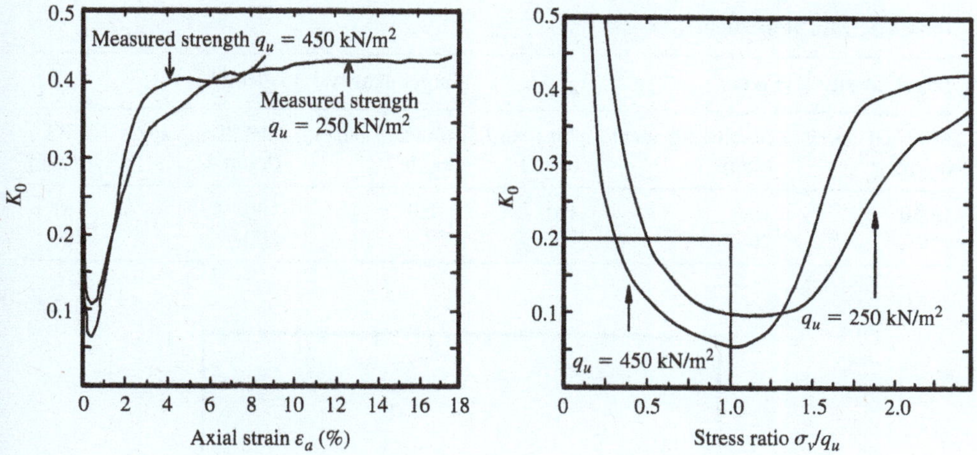

Figure 2.13. Axial strain ε_a–K_0 and axial stress ratio σ_v/q_u–K_0 relationships.

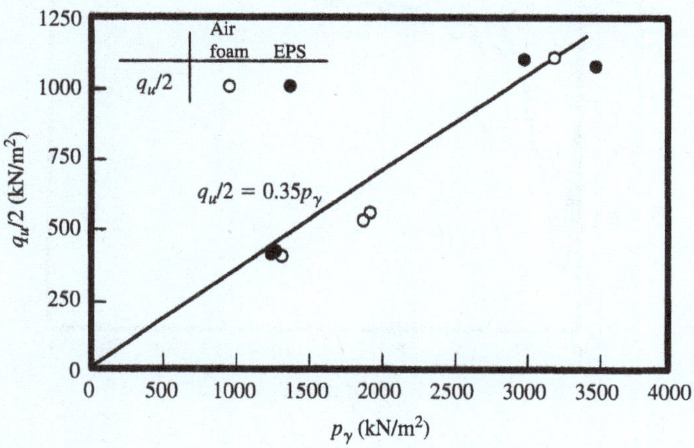

Figure 2.14. p_γ–strength relationship.

relationship obtained from the results of the K_0 consolidation test [9]. The mix proportion of the air foam treated soil used for this test is shown in Table 2.7. The clear consolidation yield stress of the air foam treated soil can be observed indicating the typical ε_a–log σ_a relationship often found in stiff clays.

Figure 2.13 shows the K_0 variations respectively with the axial strain ε_a and with the axial stress ratio [9][10]. The mix proportion of the air foam treated soil for these results is given in Table 2.7.

The figure shows that the minimum K_0 value before yield is 0.1, and it ultimately becomes constant at 0.4, but that the K_0 value at a stress ratio, σ_v/q_u of 1.0 or less is generally less than 0.2.

- Consolidation yield stress. Figure 2.14 shows the consolidation yield stresses (p_γ) obtained from the constant rate of strain consolidation test in relation with $q_u/2$ [11].

Table 2.8. Mix proportion.

Target density 1.15 g/cm^3			Target density 1.15 g/cm^3		
Adjusted slurry (kg/m^3)	Stabilizing agent (kg/m^3)	Air foam (l/m^3)	Adjusted slurry (kg/m^3)	Stabilizing agent (kg/m^3)	EPS (l/m^3)
1,050	100	161	1,050	100	161
950	200	206	950	200	206

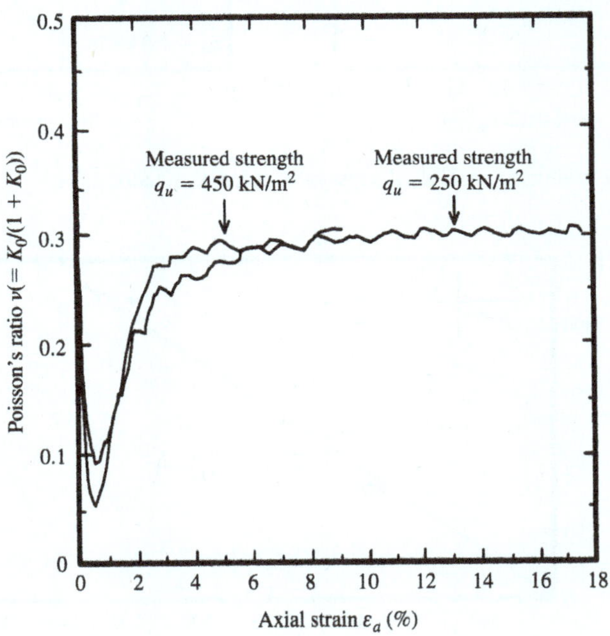

Figure 2.15. Axial strain ε_a–Poisson's ratio relationship.

Table 2.8 shows the mix proportions of the air foam treated soil. As Figure 2.14 shows, $q_u/2 \fallingdotseq 0.35p_\gamma$, confirming that this relationship is identical to that of ordinary clay ground.

- Poisson's ratio. Figure 2.15 shows Poisson's ratio obtained from K_0. Table 2.7 shows the mix proportion of the air foam treated soil. Poisson's ratio reaches its minimum value (about 0.1) at a rather smaller axial strain before yield, and it converges at about 0.3 at higher axial strain.

Triaxial compression test

The following is an example of undrained triaxial compression test results:

- Shear properties. Figure 2.16 shows examples of stress–strain curves of specimens with measured strength q_u of 250 kN/m^2 [8]. Table 2.7 shows the mix proportion of the air foam treated soil used for this test.

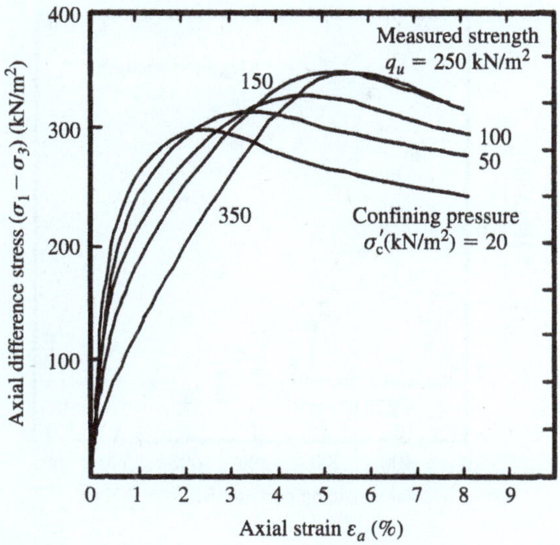

Figure 2.16. Stress–strain relationship.

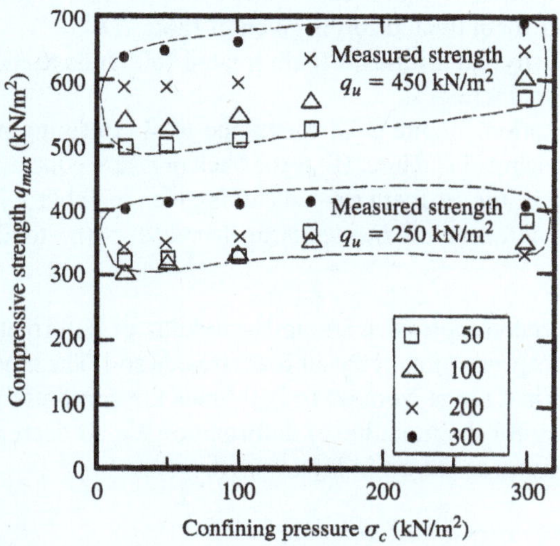

Figure 2.17. Confining pressure–compressive strength relationship.

This figure indicates that there is a clear peak in the stress–strain curve of the air foam treated soil.

- Effects of the confining pressure on the compressive strength. The air foam treated soil in a curing mold was placed in the pressure vessel filled with sea water and cured for 28 days at ambient temperature under the specified curing pressures of 50, 100, 200 and 300 kN/m^2. Figure 2.17 shows the relationship of the confining pressure with the compressive strength [8]. Note that the maximum value of the deviator stress during

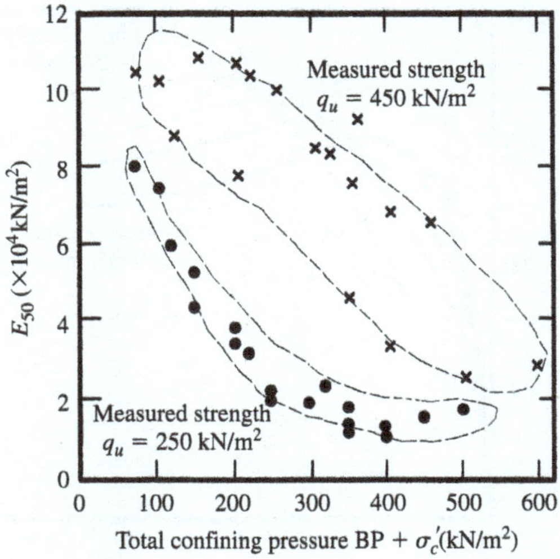

Figure 2.18. Total confining pressure–modulus of deformation relationship.

compression obtained from the stress–strain curve is $q_{max} = (\sigma_1 - \sigma_3)_{max}$. The mix proportion of the air foam treated soil is given in Table 2.7.

The compressive strength of the air foam treated soil tends to increase slightly as the confining pressure σ_c increases.

- Modulus of deformation. Figure 2.18 shows the total confining pressure–modulus of deformation relationship [11]. Here, BP is the back pressure equal to the curing pressure. The mix proportion of the air foam treated soil is given in Table 2.7.

  The modulus of deformation E_{50} tends to decrease as the total confining pressure increases.

Figure 2.19 shows the compressive strength–modulus of deformation relationship [8]. Table 2.7 shows the mix proportion of the air foam treated soil. The modulus of deformation E_{50} is distributed within a range from 40 to 260 times the compressive strength q_{max}, but this reflects a tendency for the modulus of deformation E_{50} to decrease with the increase of total confining pressure as shown in Figure 2.18.

2.2.2.6 *Dynamic properties after hardening*

Cyclic loading test

The cyclic loading test was carried out with a vibrating triaxial test apparatus. The specimens of air foam treated soil were cured for 90 days at ambient temperature and normal atmospheric pressure. Then cyclic loading was set at maximum deviator stresses, q_{max} of 0.6 q_u, 0.7 q_u, and 0.8 q_u, and the test was carried out under the lateral confining pressure of 50 kN/m^2 and the sinusoidal wave deviator stress with a frequency of 1 Hz.

Figure 2.20 shows the results of the cyclic triaxial loading test [9].

Table 2.7 shows the mix proportion of the air foam treated soil. When the loading was performed at 80% of unconfined compressive strength, failure occurred at around 1,000 cycles. However, during the loading of 60% to 70%, axial strain was only about 1.5%, even at cycles of 100,000 or more, suggesting that the soil can fully withstand loading to this extent.

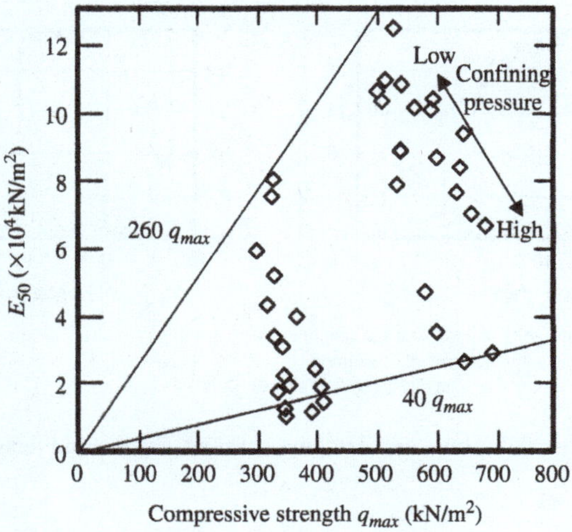

Figure 2.19. Compressive strength–modulus of deformation relationship.

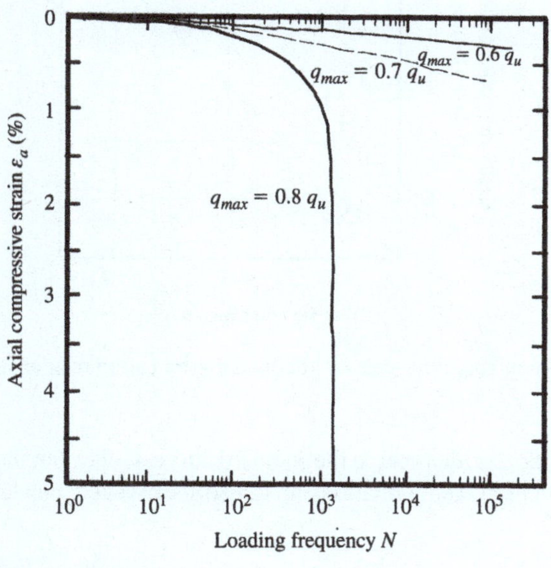

Figure 2.20. Cyclic loading test results.

Dynamic deformation properties testing

Figure 2.21 shows the results of the dynamic deformation property test (cyclic undrained triaxial test) [2].

- Target strength:$q_u = 300$ kN/m^2
- Cyclic undrained triaxial testing (confining pressure $\sigma_c' = 50$, 100 kN/m^2)
- 28-day standard curing (atmospheric pressure)

The standard curve of clay (plasticity index $I_p \geqq 30$) as a reference material is shown by the broken line. Overall, the shearing modulus ratio G/G_0–shear strain γ relationship of air

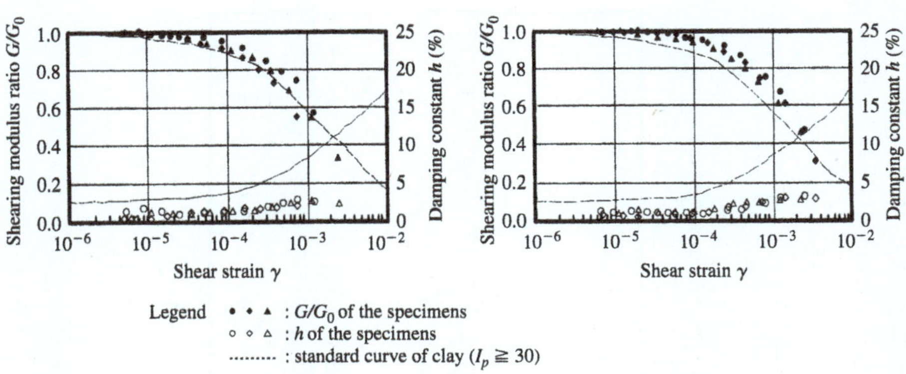

Figure 2.21. Shearing modulus ratio and damping constant–shear strain relationship.

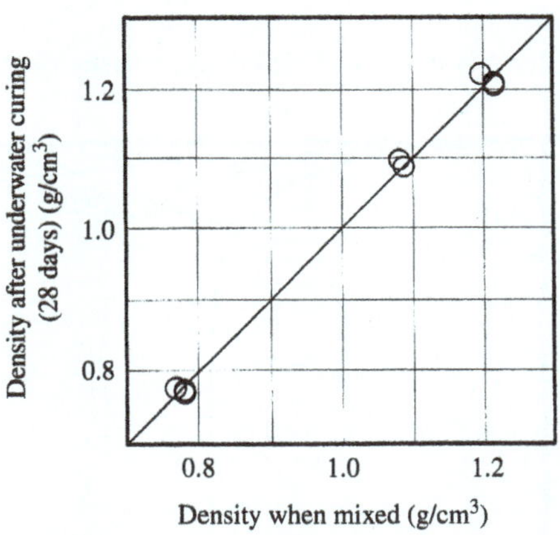

Figure 2.22. Density at mixing time–density after underwater curing relationship.

foam treated soil is almost identical to the standard curve of clay, but the damping constant h tends to decrease. The G/G_0–γ relationship is also somewhat dependent on the confining pressure.

2.2.3 Properties of beads treated soil

Beads treated soil is prepared using the materials and a preparation method according to the test conditions described in section 2.2.1, and the test method for the basic properties of samples is identical to the method used with air foam treated soil.

2.2.3.1 *Density properties*

The density of beads treated soil can be adjusted by varying the quantity of EPS beads. The higher the EPS beads content, the lower the density of the beads treated soil. Figure 2.22 compares the density of beads treated soil immediately after mixing with that after underwater curing [13]. Table 2.9 shows the mix proportion of the beads treated soil at

Table 2.9. Mix proportion.

Target density 0.8 g/cm³			Target density 1.1 g/cm³			Target density 1.2 g/cm³		
Adjustment clay (kg/m³)	Stabilizing agent (kg/m³)	EPS beads (kg/m³)	Adjustment clay (kg/m³)	Stabilizing agent (kg/m³)	EPS beads (kg/m³)	Adjustment clay (kg/m³)	Stabilizing agent (kg/m³)	EPS beads (kg/m³)
750	50	15.4	1,050	50	7.0	1,150	50	4.7
725	75	15.8	1,025	75	7.3	1,125	75	5.0
700	100	16.1	1,000	100	7.7	1,100	100	5.3

The water content of the clay is 1.8 W_L ($W_L = 76.1\%$).

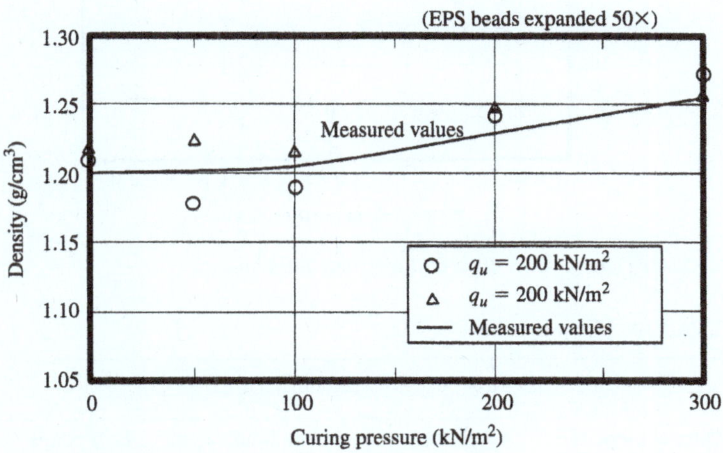

Figure 2.23. Curing pressure–specimen density relationship.

this time. This figure shows that the density when the soil is mixed and the density after underwater curing (28 days) are almost identical, and that the density is not affected by the absorption of water after underwater placing.

Figure 2.23 shows the curing pressure–density relationship of beads treated soil. The calculated densities were obtained based on the amount of compression according to the unit pressure of the EPS particles (estimated quantity). This figure reveals that the curing pressure increases with the increase of the density [14].

2.2.3.2 *Flow properties*

Figure 2.24 shows the relationship between the water content of adjusted slurry and the flow value of beads treated soil and Table 2.10 shows the mix proportion of specimens.

The flow value of the beads treated soil in the atmosphere is influenced by the water content of the adjusted slurry, so a decrease of the water content of the adjusted slurry results in a decrease of the flow value. Meanwhile, the flow value underwater is influenced by buoyancy and water pressure, so the flowability shows lower values compared to the flow values in the atmosphere.

In practical applications in the field, it is confirmed that there is no problem in feed pumping and placing of treated soils when the flow value is over 120 mm.

Figure 2.25 shows the variation of the flow value with elapsed time from mixing [13]. As the figure shows, the flowability in the atmosphere tends to decrease as time passes.

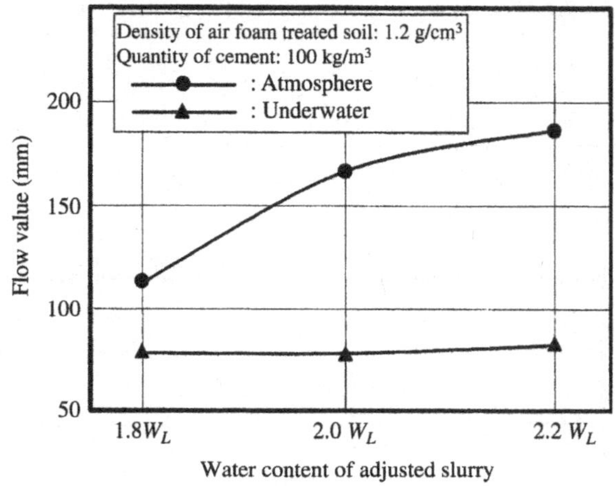

Figure 2.24. Adjusted slurry water content–flow value relationship.

Table 2.10. Mix proportion.

	Target density 1.2 g/cm^3		
Water content of adjusted slurry (%)	Adjustment clay (kg/m^3)	Stabilizing agent (kg/m^3)	EPS beads (kg/m^3)
1.8 W_L	1,100	1,100	5.3
2.0 W_L			4.5
2.2 W_L			4.0
			($W_L = 76.1\%$)

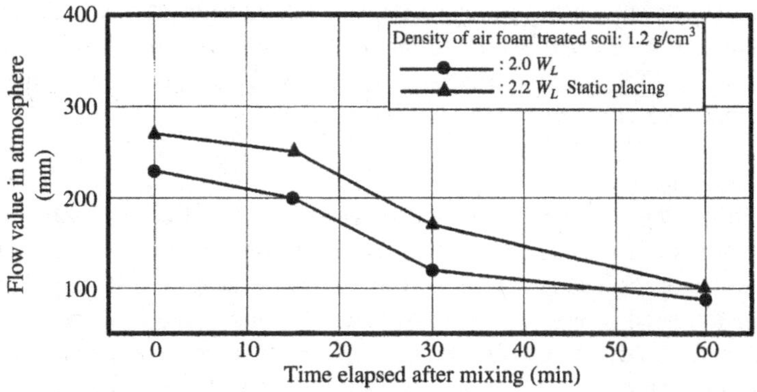

Figure 2.25. Elapsed time–flow value relationship.

2.2.3.3 *Underwater separation resistance*

The higher the water content of adjusted slurry, the more easily the constituents of beads treated soil separate (the floating of EPS beads to the surface). The separation resistance of beads treated soil is influenced by the quantity of stabilizing agent added and by the adjusted slurry. Therefore, the less stabilizing agent added and the lower the viscosity of the adjusted slurry, the more readily the materials separate (some cases of 5% separation are reported). However, as stated above, low viscosity of the adjusted slurry results in better flowability, although it increases the separation resistance. However, it is possible to select an appropriate range of the water content of the adjusted slurry that concurrently satisfies the needs for flowability and for separation resistance by performing various tests.

According to the results of tests using Kawasaki clay (marine clay) as the source soil, both high separation resistance and appropriate flowability can be attained concurrently within an adjusted slurry range of 2.0 to 2.2 W_L (W_L: liquid limit of the source soil).

It is also reported that the separation resistance varies according to the grain size and the density of the EPS beads, so the smaller their grain size and higher their density, the greater the separation resistance.

2.2.3.4 *Permeability*

As in the case of the permeability of air foam treated soil in section 2.2.2 (4), the permeability of beads treated soil is almost identical to that of the base material that contains cement.

2.2.3.5 *Strength properties*

Unconfined compression test

Figure 2.26 shows the quantity of stabilizing agent–unconfined compressive strength (28 days of aging) relationship [13] and Table 2.11 shows the mix proportion of the beads treated soil at that time.

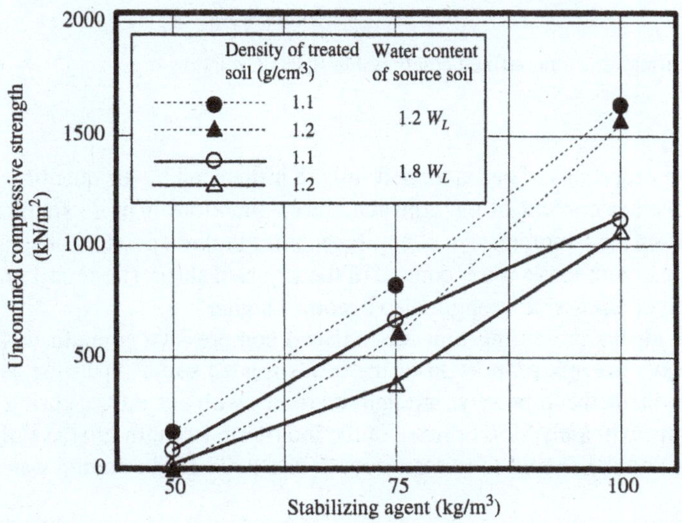

Figure 2.26. Quantity of stabilizing agent–unconfined compressive strength relationship.

Table 2.11. Mix proportion.

Target density	1.1 g/cm³			1.2 g/cm³		
Water content of adjusted slurry (%)	Adjusted slurry (kg/m³)	Stabilizing agent (kg/m³)	EPS beads (kg/m³)	Adjusted slurry (kg/m³)	Stabilizing agent (kg/m³)	EPS beads (kg/m³)
1.2 W_L	1,050	50	9.3	1,150	50	7.1
	1,025	75	9.5	1,125	75	7.4
	1,100	100	9.8	1,100	100	7.7
1.8 W_L	1,050	50	7.0	1,150	50	4.7
	1,025	75	7.3	1,125	75	5.0
	1,000	100	7.7	1,100	100	5.3

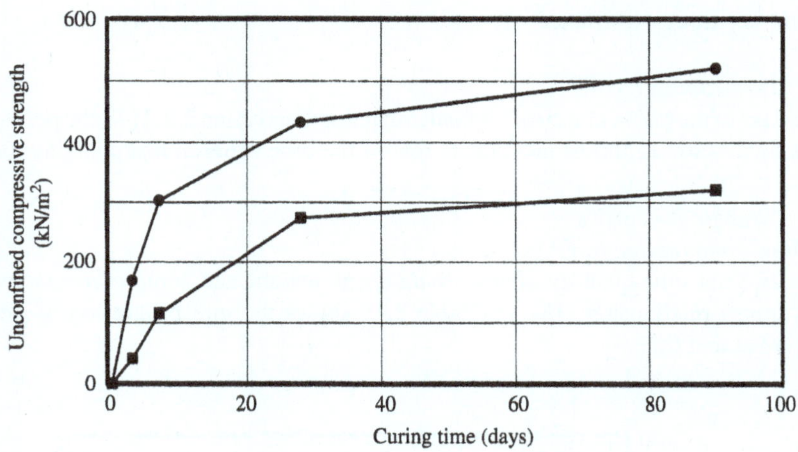

Figure 2.27. Curing time–unconfined compressive strength relationship.

The strength properties of beads treated soil are influenced by the quantity of stabilizing agent and the water content of the adjusted slurry, therefore if more stabilizing agent is added, the unconfined compressive strength becomes higher and at a constant quantity of stabilizing agent, with lower water content of the adjusted slurry (large soil grain fraction), the unconfined compressive strength also becomes higher.

Figure 2.27 shows the curing time–unconfined compressive strength relationship and Table 2.12 shows the mix proportion of the beads treated soil at that time. As this figure shows, the unconfined compressive strength increases with age during curing in sea water, and reaches approximately 80% or more of the long-term strength (90 days of aging) at 28 days of aging, and the strength increase from 7 days to 28 days of aging was $q_{u28} \fallingdotseq 1.5$ to $2.0 \cdot q_{u7}$.

Figure 2.28 shows the curing pressure–unconfined compressive strength relationship [14], and Table 2.12 shows the mix proportion of the beads treated soil for this test.

Table 2.12. Mix proportion.

	Target density 1.2 g/cm^3		
Water content of adjusted slurry (%)	Adjusted slurry (kg/m^3)	Stabilizing agent (kg/m^3)	EPS (kg/m^3)
2.2 W_L	1,136 1,127	64 73	3.2 3.3

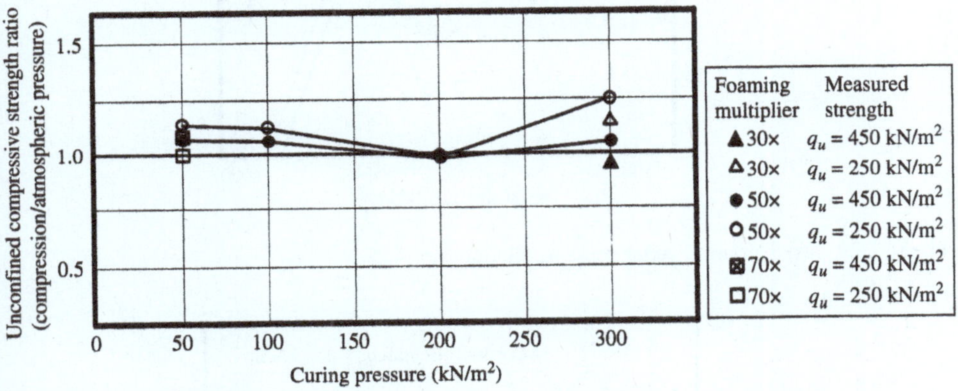

Figure 2.28. Curing pressure–unconfined compressive strength relationship.

The strength of a specimen with pressurized curing is almost identical to that with atmospheric curing. Therefore, the curing pressure has little effect on strength development, and the effect of differences in curing pressure can also be neglected.

One-dimensional consolidation test
Figure 2.29 summarizes the axial strain ε_a–axial stress σ_a relationship obtained from the results of the K_0 consolidation testing [8]. Table 2.12 shows the mix proportion of beads treated soil for this test.

From the figure, the consolidation yield stress of the beads treated soil clearly appears, and the ε_a–log σ_a relationship is similar to that of stiff clays.

- Coefficient of earth pressure at rest. K_0 consolidation testing was carried out using a double cell type triaxial apparatus. After initial isotropic consolidation of the specimen at the pressure of 20 kN/m^2, by setting the rate of increase of axial pressure at 65 kN/m^2/hr, while controlling the cell pressure to prevent lateral strain, the axial pressure was increased up to 1,000 kN/m^2. Figure 2.30 shows the variation of K_0 with axial strain [8]. Table 2.12 shows the mix proportion of the beads treated soil for this test. The minimum K_0 value ranges from 0.1 to 0.2 before yield, but if the axial strain is increased, it ultimately converges at 0.4.
- Consolidation yield stress. The *e*-log*p* curve of the beads treated soil obtained by the constant rate of strain consolidation test is almost identical to the relationship for clayey

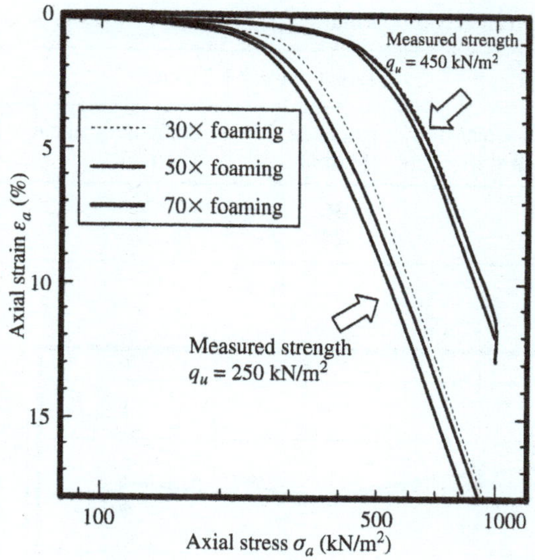

Figure 2.29. Axial strain ε_a–axial stress σ_a relationship.

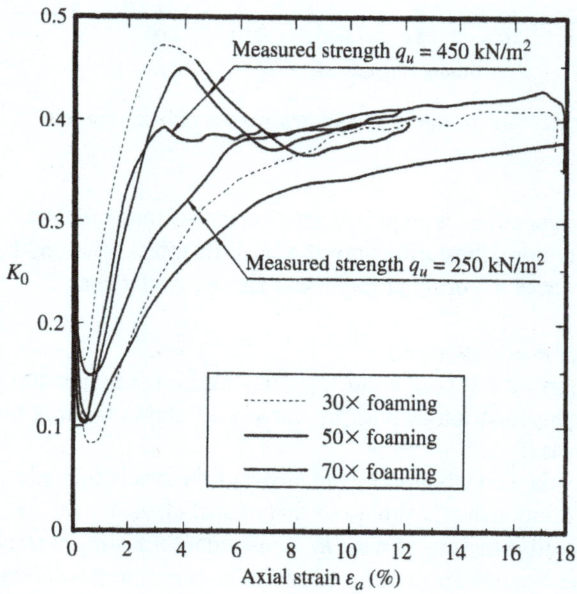

Figure 2.30. Axial strain ε_a–K_0 relationship.

soil, and it is possible to obtain the clear consolidation yield stress p_y. It was confirmed that there is a close relationship between the consolidation yield stress p_y and strength $(q_u/2)$ that is almost identical to that with ordinary clay ground as shown in the equation below [11].

$$(q_u/2) \fallingdotseq 0.35 \cdot p_y$$

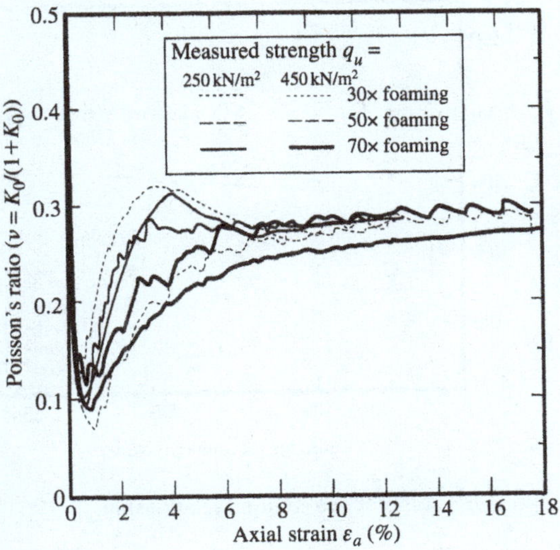

Figure 2.31. Axial strain ε_a–Poisson's ratio relationship.

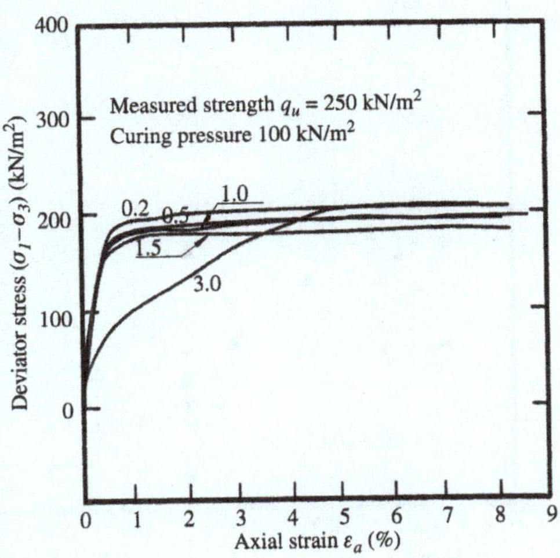

Figure 2.32. Stress–strain curve.

- Poisson's ratio. Figure 2.31 shows the axial strain–Poisson's ratio relationship obtained by calculation based on K_0 consolidation tests [8]. Table 2.12 shows the mix proportion of beads treated soil. Poisson's ratio of beads treated soil reaches its minimum value (0.1 to 0.15), but as the axial strain increases, the ratio converges at about 0.3 at a smaller strain before yielding.

Triaxial compression test (undrained)

- Properties of shear. Figure 2.32 shows examples of the stress–strain curves of specimens with a measured strength of 250 kN/m^2 [9]. Table 2.12 shows the mix proportion of beads treated soil for this test.

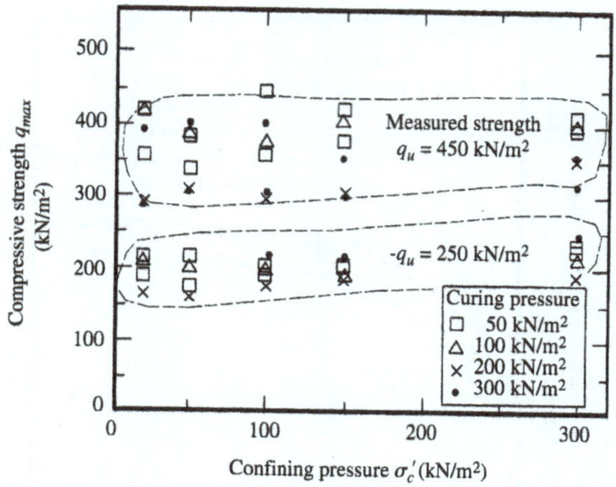

Figure 2.33. Confining pressure–compressive strength relationship.

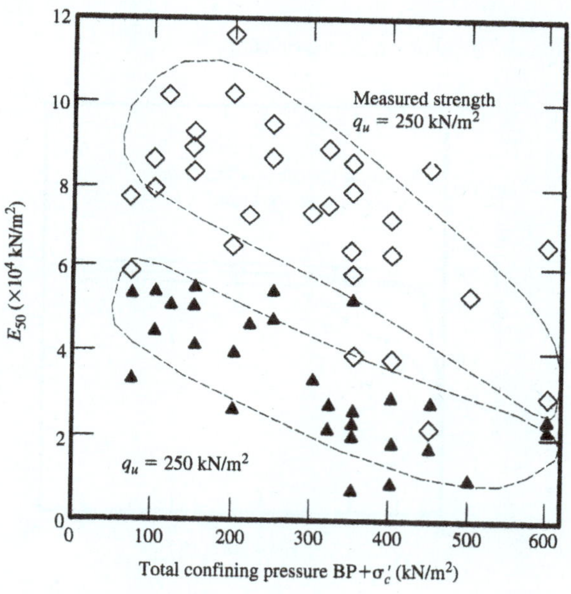

Figure 2.34. Total confining pressure–E_{50} relationship.

This figure shows that there is no clear peak value of stress for beads treated soil.

- Effects of the confining pressure and curing pressure on the compressive strength. EPS beads with 50× expansion were used for the beads treated soil. The specified curing of the specimens was carried out by placing the specimens in a pressurized vessel filled with sea water in a static manner.

Specimens were cured for 28 days at room temperature under increasing confining pressures (50, 100, 200 and 300 kN/m²). Figure 2.33 shows the confining pressure–compressive strength relationship [7]. Table 2.12 shows the mix proportion of beads treated soil for this test.

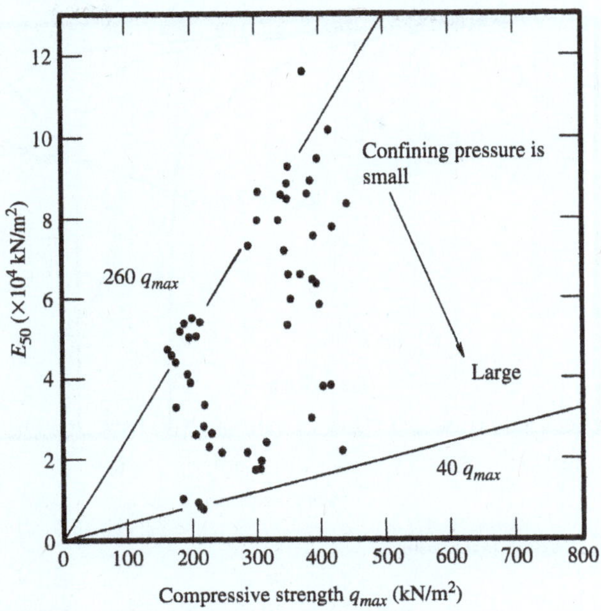

Figure 2.35. Compressive strength–E_{50} relationship.

It can be concluded that the compressive strength of beads treated soil is almost unaffected by the confining pressure σ'_c.

- Modulus of deformation. Figure 2.34 shows the total confining pressure–modulus of deformation relationship and Figure 2.35 shows the compressive strength–modulus of deformation relationship [7]. Table 2.12 shows the mix proportion of beads treated soil.

  The modulus of deformation E_{50} tends to decrease as the total confining pressure increases, and is distributed in a range of 40 to 260 times the value of q_{max}.

2.2.3.6 *Dynamic properties after hardening*

Cyclic loading test

The cyclic loading test was carried out with a vibrating triaxial test apparatus. The specimens of beads treated soil were cured for 90 days at ambient temperature and normal atmospheric pressure, then cyclic loading was set at maximum deviator stresses, q_{max} determined as 0.6 q_u, 0.7 q_u, and 0.8 q_u, and the test was carried out under the lateral confining pressure of 50 kN/m^2 and the sinusoidal wave deviator stress with a frequency of 1 Hz. Figure 2.36 shows the results of the cyclic loading test [8]. Table 2.12 shows the mix proportion of the beads treated soil for this test. When it is loaded at 80% of unconfined compressive strength, failure occurred at around 1,000 cycles, but during the loading of 60%, strain was only about 1.5%, even at cycles of 100,000 or more, suggesting that the soil can fully withstand loading to this extent.

Dynamic deformation properties testing

Figure 2.37 shows the results of the dynamic deformation property test [11]. Table 2.12 shows the mix proportion of the beads treated soil. The test conditions are shown in next page.

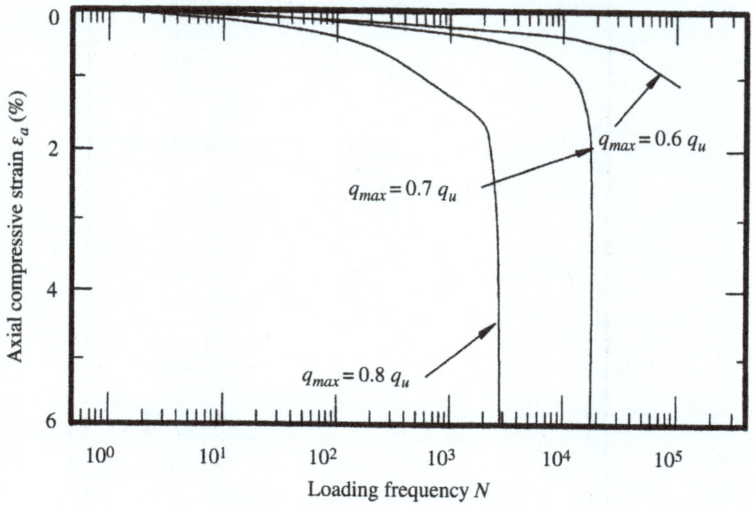

Figure 2.36. Cyclic loading test results.

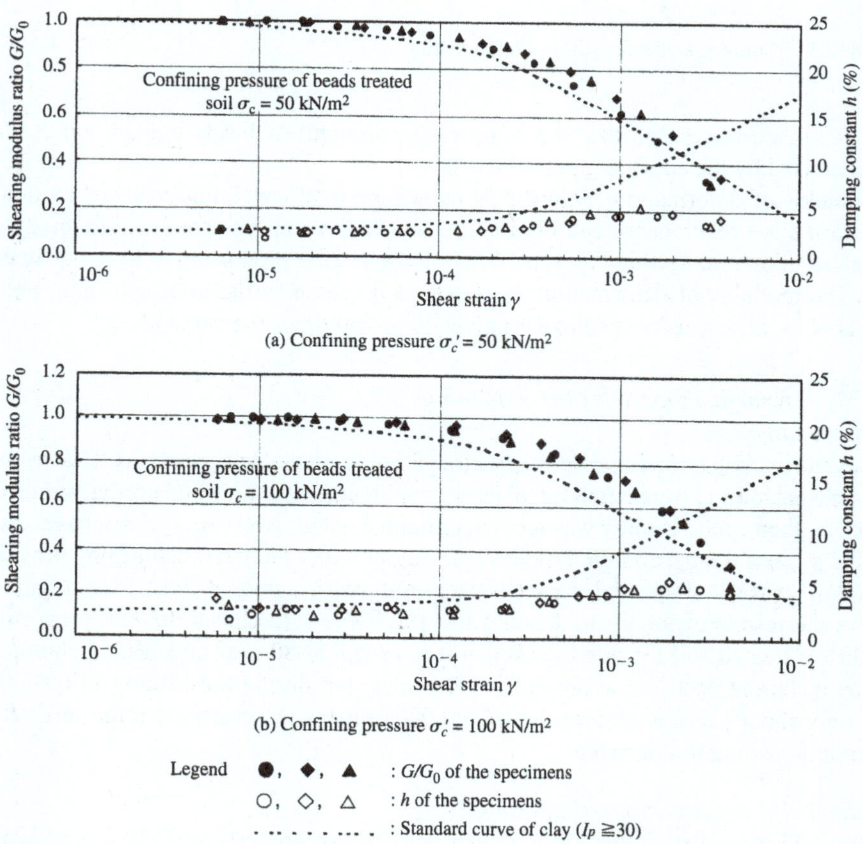

Figure 2.37. Shearing modulus ratio and damping constant–shear strain relationship.

- Target strength $q_u = 300\,\text{kN/m}^2$
- Cyclic undrained triaxial testing (confining pressure $\sigma_c' = 50, 100\,\text{kN/m}^2$)
- 28-day standard curing (atmospheric pressure)

The standard curve of clay (plasticity index $I_p \geqq 30$) as a reference material is shown by the broken line. Overall, the shearing modulus ratio G/G_0 with the shear strain of beads treated soil is comparatively low. It also shows that the G/G_0–γ relationship is slightly dependent on the confining pressure.

REFERENCES

1. Coastal Development Institute of Technology: Technical Manual of SGM Lightweight Soil Method, 1999.
2. M. Tanaka, D. Takeuchi, H. Shinsha, K. Sasaki, M. Yoshihara: Examination connected with Density Adjustment of Lightweight Geo-Material Made of Dredged Slurry mixed with Air Foam, Proceedings of the 31st Conference on Geotechnical Engineering, pp. 2491–2492, 1996.
3. T. Wako, Y. Matsunaga, D. Takeuchi, T. Fukasawa, T. Kishida: Drying properties of lightweight treated soil, Fifty-second Annual Technical Conference of the Japan Society of Civil Engineers, 1997.
4. J. Kasai, T. Tsuchida, J. Mizukami, Y. Yokoyama, K. Tsuchida, T. Goto: Effects of Curing Condition for Lightweight Materials, Proceedings of the 28th Conference on Soil Engineering, pp. 2669–2672, 1993.
5. T. Tsuchida, K. Nagai, T. Okumura, T. Kishida, K. Funada: Mechanical Properties of Lightweight Geo-Material Used for the Backfill of Quaywalls (Part 1), Proceedings of the 31st Conference on Geotechnical Engineering, pp. 2525–2528, 1995.
6. K. Ishizuka, T. Okumura, H. Kuroyama, T. Hori, K. Ishitani: Properties of Lightweight Geo-Material made of Dredged Slurry mixed with Air Foam (Part 1) Mix Proportion Study, Proceedings of the 29th Conference on Soil Engineering, pp. 2415–2416, 1994.
7. Y. Kikuchi, H. Yoshino: Permeability of Lightweight Soil Made of Dredged Slurry Mixed with Air Foam, Report of the Port and Harbor Research Institute, Vol. 37, No. 1, 1998.
8. Y. X. Tang, T. Tsuchida, A. Shirai, H. Ogata, K. Shiozaki: Triaxial Compression Characteristics of Super Geo-Material Cured Underwater, Proceedings of the 31st Conference on Geotechnical Engineering, pp. 2493–2494, 1996.
9. Y. X. Tang, T. Okumura, K. Ishitani, M. Kagamida, T. Bessho: Coefficient of Earth Pressure at Rest and Enduring Property under Repeated Load for Super Geo-Material, Proceedings of the 31st Conference on Geotechnical Engineering, pp. 2495–2496, 1996.
10. Japan Port and Harbor Association: Technical standards and commentaries for harbor facility technologies, Revised Edition, Vol. 1, 1999.
11. T. Tsuchida, Y. Yokoyama, J. Mizukami: Field Test of Lightweight Geomaterials for Harbor Structures, Technical Note of the Port and Harbor Research Institute, No. 833, 1996.
12. Y. Kikuchi, D. Takeuchi, T. Chiyoda, M. Kagamita, A. Shirai: Dynamic Characteristics of Lightweight Geo-Material made of Dredged, Proceedings of the 32nd Geotechnical Engineering Conference, pp. 2575–2576, 1997.
13. T. Tsuchida, M. Yukawa, M. Kagamida, I. Kitamori, S. Yamada: Properties of Lightweight Geo-material made of Dredged Slurry mixed with Expanded Resin Beads (Part 1) Mix Proportion Study, Proceedings of the 29th Conference on Soil Engineering, pp. 2385–2386, 1994.
14. Y. Kikuchi, Y. Umehara, M. Uchiyama, T. Gotou, H. Ogata: Properties of Lightweight Geo-material made of Dredged Slurry mixed with Expanded Resin Beads (Part 1) Effect of Curing Condition for Lightweight Geo-material, Proceedings of the 29th Conference on Soil Engineering, pp. 2391–2392, 1994.

15. T. Tsuchida, T. Wako, H. Matsushita, M. Yoshiwara: Evaluation of Washout Resistance of Lightweight Treated Soil Cast Underwater, Technical Note of the Port and Harbor Research Institute, No. 884, 1997.
16. T. Tsuchida, H. Matsushita, Y. Omodaka: Development of Washout Resistance Test of Casting Velocity Control Type of Lightweight Treated Soil Cast Underwater (Part 1), Proceedings of the 33rd Conference on Geotechnical Engineering, pp. 2455–2456, 1998.
17. T. Tsuchida, T. Wako, H. Matsushita, M. Yoshihara: Washout Resistance Test of Casting Velocity Control Type of Lightweight Treated Soil Cast Underwater (Part 2), Proceedings of the 33rd Conference on Geotechnical Engineering, pp. 2457–2458, 1998.
18. T. Tsuchida: Development and Use of Foamed Treated Soil in Port and Airport Project, Report of the Port and Harbor Research Institute, Vol. 38, No. 2, 1999.
19. T. Satoh, T. Tsuchida, K. Mitsukuri, Z. Hong: Field placing test of lightweight treated soil under sea water in Kumamoto Port, *Soils and Foundations*, Vol. 41, No. 5, pp. 145–154, 2001.
20. T. Satoh, N. Ueno, K. Mitsukuri, K. Kawano, T. Tsuchida: Under casting test of lightweight treated soil made of waste soils, Coastal Geotechnical Engineering in Practice, Proceedings of IS-Yokohama 2000, Vol. 1, pp. 709–714, 2000.9.
21. M. Hirasawa, S. Saeki, S. Kodama, T. Yakuwa, T. Tsuchida: Development of light-weight soil using excavated sand and its application for harbor structures in cold regions, Coastal Geotechnical Engineering in Practice, Proceedings of IS-Yokohama 2000, Vol. 1, pp. 559–604, 2000.9.
22. T. Tsuchida, T. Okumura and T. Kishida: Use of Artificial Lightweight Materials for Backfilling of Quaywalls, 2nd International Conference on Soft Soil Engineering, Nanjin, Vol. 2, pp. 807–812, 1996.5.
23. T. Tsuchida, T. Okumura, D. Takeuchi, T. Kishida: Development of Lightweight Fill from Dredging, 2nd International Congress on Environmental Geotechnics, Osaka, Vol. 1, pp. 415–420, 1996.11.

Chapter 3

Design and Sample Design

3.1 OUTLINE

3.1.1 Basic design concept

The addition of a stabilizing agent increases the strength of lightweight treated soil. However, because this strength is far lower than the strength of structural concrete materials, it is likely to be appropriate to classify the lightweight treated soil as a soil. It is, therefore, assumed that the properties of lightweight treated ground are almost equivalent to the ground of stiff clays except for its lightweight characteristics. Therefore, the designing of lightweight treated ground is carried out using the same method and procedures as clay ground. Details of the various design parameters used for lightweight treated ground are presented in section 3.2.

3.1.2 Earth pressure of lightweight treated ground

In a case where the lightweight treated ground is semi-infinite with constant thickness of soil layer running parallel horizontally, the earth pressure can be calculated using Coulomb's equation. In this case, the lightweight treated ground is treated as clayey soil.

The scope of application of ground improvement such as the lightweight treated soil method is restricted to finite conditions. Therefore, it is difficult to evaluate the earth pressure reduction effects appropriately using conventional earth pressure equations. Therefore, to evaluate the active earth pressure for lightweight treated ground, the slice method which can consider arbitrary dimensions is used, as described in Section 3.3.

3.1.3 Liquefaction of lightweight treated ground

The addition of a stabilizing agent to lightweight treated soil provides it with adequate strength (cohesiveness). Therefore, it is not necessary to study the liquefaction of lightweight treated ground.

3.2 SOIL PARAMETERS OF LIGHTWEIGHT TREATED SOIL

The parameters to be used to design lightweight treated soil are established accounting for application method, curing conditions, etc. by appropriate investigation and laboratory tests for the source and treated soils.

In principle, the soil parameters used to design lightweight treated soil are evaluated by laboratory tests which model the environmental conditions of the site or by other methods.

When it is difficult to carry out such tests, the soil parameters can be calculated based on the following concepts with reference to the test results presented in chapter 2.

3.2.1 Determination of the unit weight

The unit weight of lightweight treated soil can be adjusted within a range $\gamma_t = 6$ to 15 kN/m^3 (density of 0.6 to 1.5 g/cm^3) by varying the mix proportion ratio or water added to the mixture. However, it is difficult to place lightweight treated soil underwater when it is lighter than the unit weight of sea water or fresh water. If the water level or ground water level is likely to rise after placing, causes the ground nearing, the design must be considered carefully.

The unit weight of lightweight treated soil may show a slight increase as it absorbs water over time. Therefore, the design should be carried out while considering this extra increase in the unit weight of about 0.5 kN/m^3 (density of 0.05 g/cm^3).

3.2.2 Determination of the design strength

The strength of lightweight treated soil is to be determined to secure the stability of the structure and the treated ground under external forces acting upon them.

The static strength of lightweight treated soil is solidification strength developed by cement stabilizers. The static strength of lightweight treated soil is generally evaluated by the unconfined compressive strength q_u. This unconfined compressive strength can be determined freely within a range of approximately 100 to 500 kN/m^2. The strength of this material can be treated as a cohesive material with $c =$ constant ($\phi = 0$).

3.2.3 Shear strength used for design

Shear strength c_u used for design is the undrained shear strength.

$$c_u = q_u/2$$

3.2.4 Modulus of deformation, E_{50}

If test data are available, the modulus of deformation E_{50} can be obtained from them. If not, E_{50} is estimated by the following equation based on the unconfined compressive strength q_u.

$$E_{50} = 100 \text{ to } 200\, q_u$$

where, E_{50}: modulus of deformation (kN/m^2); q_u: unconfined compressive strength (kN/m^2).

(The above modulus of deformation is a value that corresponds to strain levels from 0.05 to 0.20%.)

3.2.5 Poisson's ratio

The following values are used as Poisson's ratio in accordance with past measurements:

- for air foam treated soil, $\nu = 0.05$ to 0.15
- for beads treated soil, $\nu = 0.10$ to 0.20

(The above values are those within a strain range from 1.0 to 2.0% at or below the consolidation yield stress.)

3.2.6 Coefficient of friction

In earth pressure calculations and stability calculations against sliding, the following coefficients of friction are used for design at present:

- for lightweight treated soil and sand, $\mu = 0.55$ to 0.60
- for lightweight treated soil and riprap, $\mu = 0.75$ to 0.80

3.3 ACTIVE EARTH PRESSURE OF LIGHTWEIGHT TREATED GROUND

3.3.1 Active earth pressure calculation method

In conventional earth pressure calculations, it is assumed that the ground is semi-infinite, has constant thickness, and is deposited horizontally. However, it is difficult to use those conventional earth pressure equations to correctly evaluate the earth pressure reduction effects, when lightweight treated soil is used as backfill for a structure in order to reduce the earth pressure acting on a wall body. In such cases, therefore, the earth pressure is calculated using the analysis method that can appropriately account for the finite shape of backfill.

For the above reasons, earth pressure acting on a wall body is calculated by the slice method that can appropriately account for a finite shape during both normal and earthquake conditions.

3.3.2 Calculation of the earth pressure resultant force by the slice method

The earth pressure resultant force is calculated by the slice method as follows. First, the surface behind the wall body structure is assumed. Then, the block of soil between the slip surface and the wall surface is sliced. The earth pressure resultant force is calculated based on equilibrium of the weight, buoyancy, shear force on the slip surface, and seismic force on each slice (see Figure 3.1).

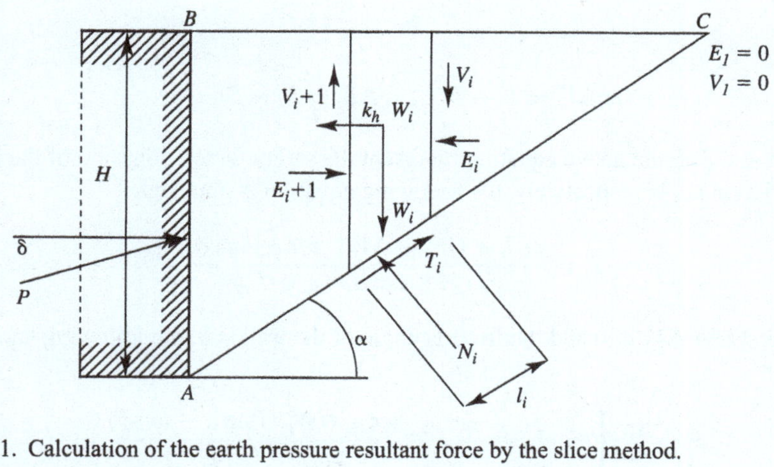

Figure 3.1. Calculation of the earth pressure resultant force by the slice method.

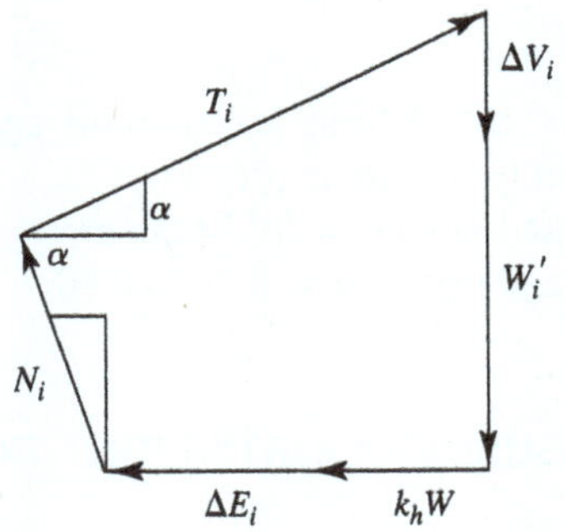

Figure 3.2. Equilibrium of the forces for each slice.

where,

P : earth pressure resultant force
α : slip surface angle
W_i : total mass of the slice
W_i' : effective mass of the slice
T_i : shearing force of the slip surface
V_i : vertical force acting on the right side of the slice
k_h : lateral seismic coefficient
δ : angle of friction of the wall
l_i : length of the slip surface of the slice
N_i : vertical force of the slip surface
E_i : lateral force acting on the right side of the slice

The equilibrium of these forces is shown in Figure 3.2. The vertical and lateral equilibrium is established by the following equations:

$$W_i' + \Delta V_i = T_i \cdot \sin\alpha + N_i \cdot \cos\alpha$$
$$W_i \cdot k_h + \Delta E_i = T_i \cdot \cos\alpha - N_i \cdot \sin\alpha$$

where,

$$\Delta V_i = V_i - V_{i+1}, \quad \Delta E_i = E_i - E_{i+1}$$

If F_s, c, and ϕ in the above equation represent the safety factor, cohesion of the ground, and the friction angle respectively, the following equation is obtained.

$$T_i = \frac{c \cdot l_i + (W_i' + \Delta V_i) \cdot \sec\alpha \cdot \tan\phi}{F_s \cdot (1 + \tan\alpha \cdot \tan\phi / F_s)}$$

Assuming that $\Delta V_i = 0$ and the friction angle of the wall is 0, the following equation is obtained.

$$P = \sum \left\{ W_i \cdot k_h + \frac{-c \cdot l_i \cdot \sec\alpha + W_i' \cdot (\tan\alpha - \tan\phi)}{1 + \tan\alpha - \tan\phi} \right\}$$

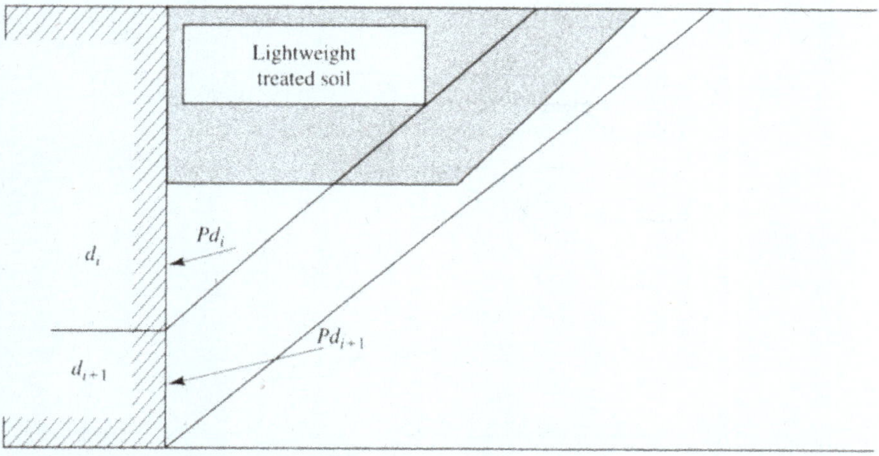

Figure 3.3. Earth pressure in the slice method.

It is possible to use this equation to calculate the earth pressure of the slip surface that is assumed behind the structure passing through point A in Figure 3.1. The values used for actual calculation are obtained by varying α to find P, then taking its maximum value as the active earth pressure resultant force and α at that time as the failure angle.

3.3.3 Calculation of the earth pressure strength by the slice method

The earth pressure strength p is calculated as shown below.

The earth pressure resultant force at each depth of the wall surface is calculated as shown in Figure 3.3. The earth pressure strength p is calculated by dividing the difference between the earth pressure resultant forces of adjoining layers by the depth difference.

$$p = \frac{(\text{total earth pressure at depth } d_{i+1}; Pd_{i+1}) - (\text{total earth pressure at depth } d_i; Pd_i)}{d_{i+1} - d_i}$$

3.3.4 Failure mode of the earth pressure calculation based on the slice method

The following three cases of failure modes of the earth pressure calculation based on the slice method are assumed for simplification. The active earth pressure is the maximum value obtained by calculating the earth pressure for each of the following three failure modes, and α at that time is the failure angle.

The three failure modes are:

- Mode I: linear slip cutting the interior of the lightweight treated soil;
- Mode II: slip considering a hypothetical crack in the lightweight treated soil; (slip in the case where the shear resistance on the vertical surface of the interior of the lightweight treated soil is not considered)
- Mode III: slip occurrence along the boundary of the lightweight treated soil.

For the shear resistance of the boundary surface, the coefficient of friction in section 3.2 (6) is used.

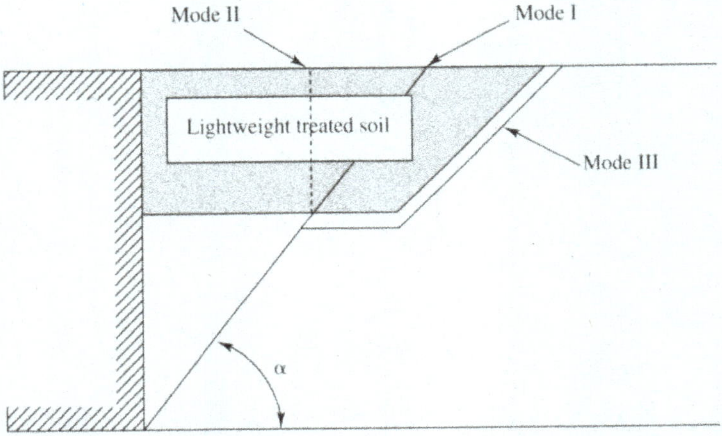

Figure 3.4. Three failure modes when a slip surface passes through lightweight treated soil.

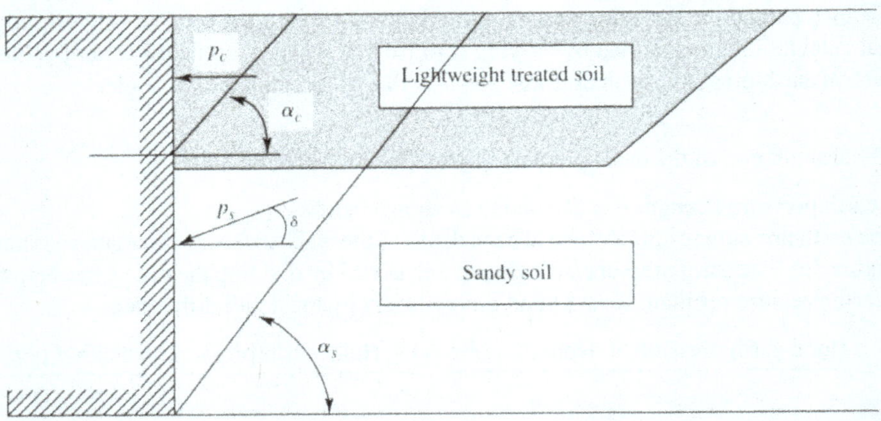

Figure 3.5. Concept of wall surface friction angle.

3.3.5 Wall surface friction angle

There are many unclear points regarding the wall surface friction angle of lightweight treated ground. Therefore, the wall surface friction angle refers to the values under present conditions.

- Clayey soil (including the lightweight treated soil): the viscosity acting between the soil and the wall surface can be neglected.
- Sandy soil: the wall surface friction angle, δ is 0.26 rad (15°) is used as the standard.

In the earth pressure calculation based on the slice method, the value of the wall surface friction angle is adopted for the soil layer at the origin point (wall surface side) of the slip surface.

3.3.6 Earth pressure during placing

The earth pressure during placing varies according to the viscosity of the lightweight treated soil and the placing speed. Because the pouring lift so far in the field has been about 1.0 m, in the present design method, the lightweight treated soil placed is treated as a fluid body to calculate the earth pressure as lateral pressure proportional to the density and the depth of this fluid body.

When lightweight treated soil is placed in a number of separate layers, each layer is placed after the required strength of the lower layer has developed.

3.4 EXAMPLES OF THE DESIGN OF A QUAYWALL IN A CASE WHERE LIGHTWEIGHT TREATED SOIL IS USED AS THE BACKFILL

3.4.1 Outline

In these design examples, the slice method is used to calculate the active earth pressure in a case where lightweight treated soil is used. The following two kinds of cases were studied. Earth pressure and the section were compared in a case where lightweight treated soil was used and in a case where it was not.

In the case of the steel sheet pile quaywall, only the design procedure, design conditions, analysis results, section diagram, and earth pressure diagram are introduced.

- Example of design using lightweight treated soil behind a caisson type quaywall
- Example of design using lightweight treated soil behind a steel sheet pile type quaywall

3.4.2 Example of design using lightweight treated soil behind a caisson type quaywall

Design procedure
The quay was designed according to the following procedure.

Design conditions
- Design conditions

Tide level	H.W.L. : D.L. +1.5 m L.W.L. : D.L. ±0.0 m R.W.L. : D.L. +0.5 m
Planned water depth	D.L. −10.0 m
Design water depth	D.L. −10.5 m
Crest height	Superstructure crest height: D.L. +3.0 m Caisson crest height: D.L. +2.0 m
Overburden load	Normal: $w = 20\,\text{kN/m}^2$ Earthquake: $w' = 10\,\text{kN/m}^2$
Design earthquake intensity	Aerial seismic coefficient: 0.15
Coefficient of friction	Concrete with riprap: 0.60

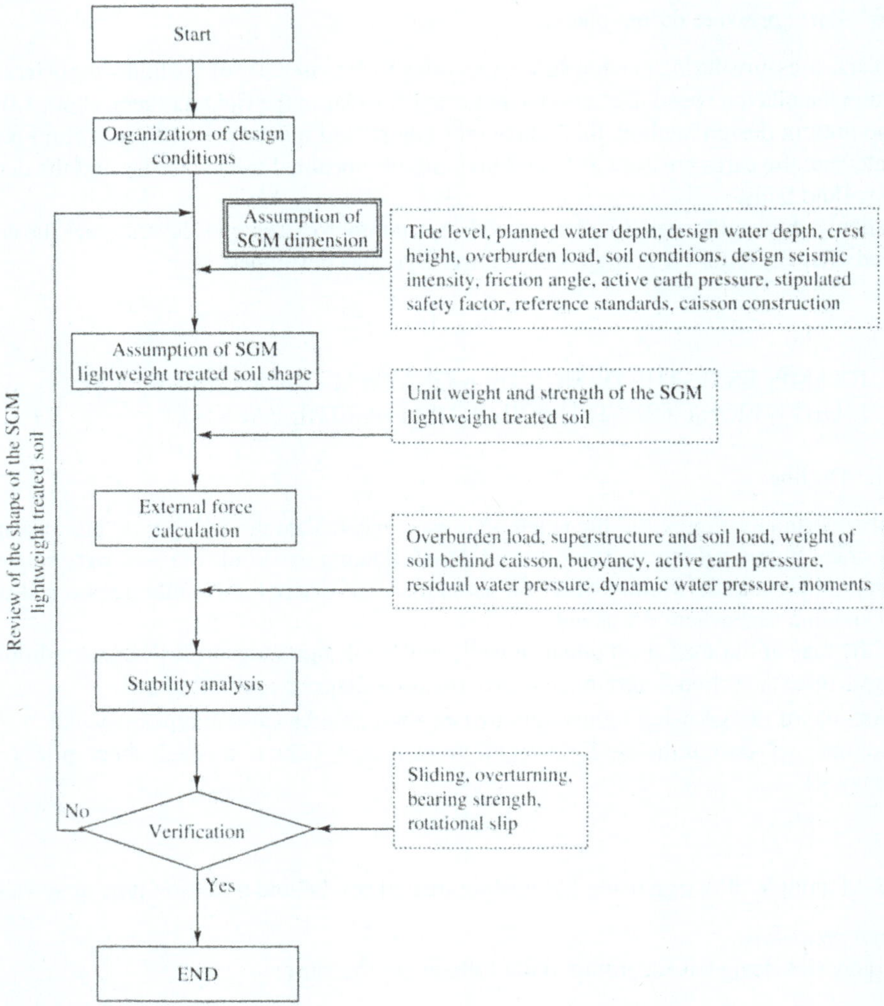

Figure 3.6. Design procedure flow chart.

- Soil conditions

Lightweight treated soil	Weight in the atmosphere: 10.0 kN/m^3 Weight underwater: 2.0 kN/m^3 Viscosity: 100.0 kN/m^3
Backfill rock	Weight in the atmosphere: 18.0 kN/m^3 Weight underwater: 10.0 kN/m^3 Internal friction angle: 0.70 rad (40°)
Reclamation soil	Weight in the atmosphere: 18.0 kN/m^3 Weight underwater: 10.0 kN/m^3 Internal friction angle: 0.52 rad (30°)

(*contd.*)

Foundation ground	Top layer (D.L. −10.0 m to D.L. −20.0 m) Weight in the atmosphere: 18.0 kN/m^3 Weight underwater: 10.0 kN/m^3 Internal friction angle: 0.52 rad (30°) Bottom layer (D.L. −20.0 m and deeper) Weight in the atmosphere: 18.0 kN/m^3 Weight underwater: 10.0 kN/m^3 Internal friction angle: 0.61 rad (35°)
Coefficient of friction of lightweight treated soil with riprap: 0.75	
Coefficient of friction of lightweight treated soil with sand: 0.60	

- Earth pressure calculation conditions (active earth pressure)

Normal condition	Earth pressure by slice method
During earthquake	Earth pressure by slice method

- Allowable safety factor

Study item	Normal condition	During earthquake
Sliding	1.2	1.0
Overturning	1.2	1.1
Rotational slip	1.3	
Toe pressure (kN/m^2)	450.00	450.00

- Study section diagram. The study section is shown in Figure 3.7.
- Caisson construction diagram. The caisson construction diagram is shown in Figure 3.8.

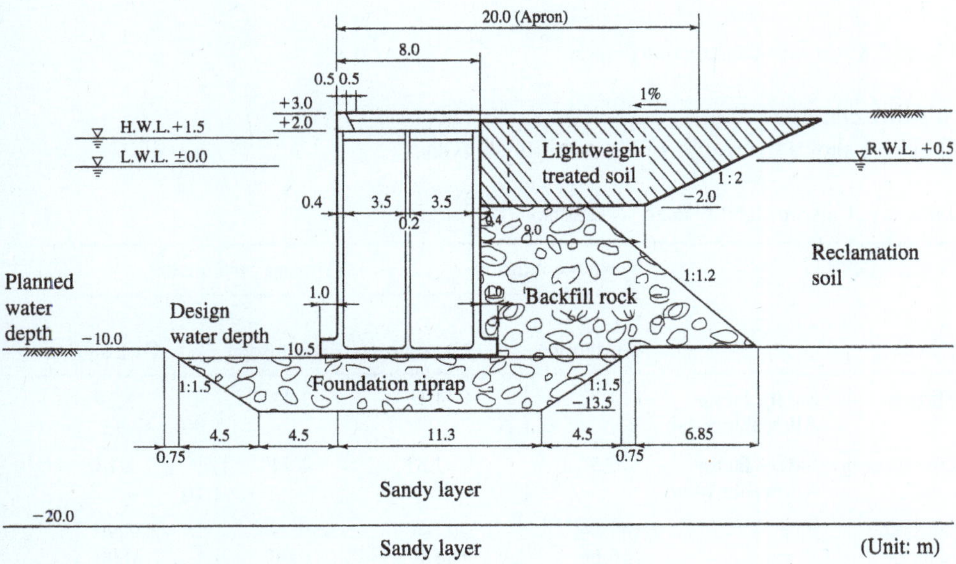

Figure 3.7. Section diagram.

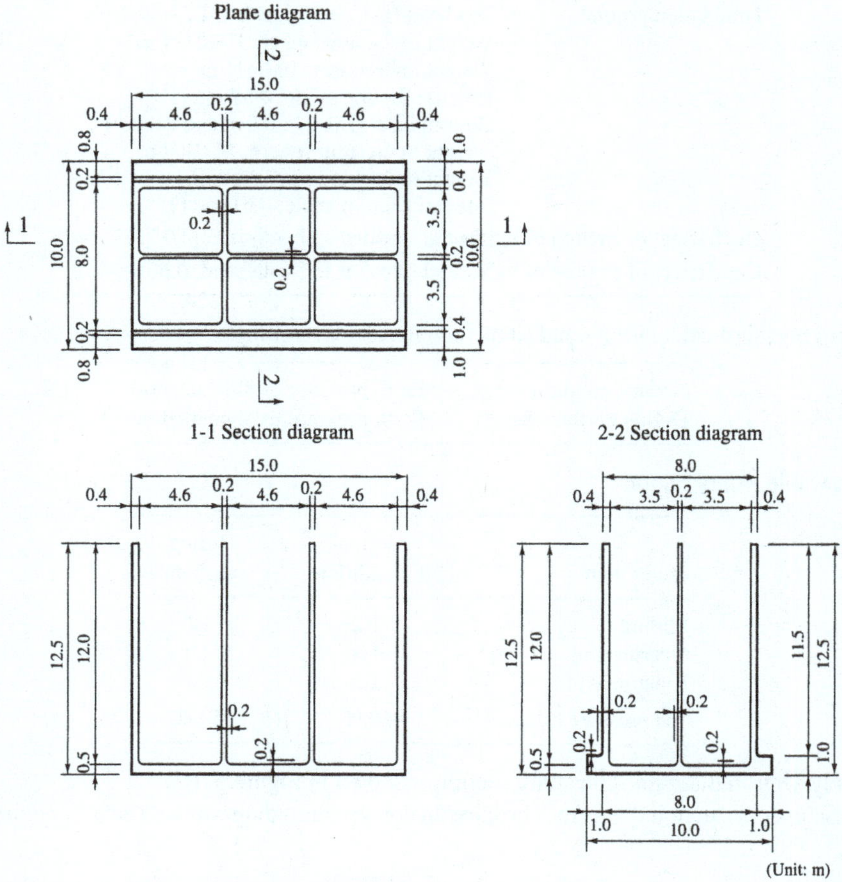

Figure 3.8. Caisson construction diagram.

Stability analysis

Table 3.1 shows the results of the stability analysis.

Table 3.1. Caisson stability analysis results.

Category		Normal condition		During earthquake	
		No overburden	Overburden	No overburden	Overburden
Sliding	Safety factor	3.72	4.16	1.13	1.17
	Allowable value	1.20		1.00	
Overturning	Safety factor	6.56	7.33	2.11	2.14
	Allowable value	1.20		1.10	
Subgrade reaction	$p_1 =$	189.86	202.40	359.29	375.96
	$p_2 =$	116.66	140.12	0.00	0.00

(contd.)

Table 3.1. (*contd.*)

Category		Normal condition: No overburden	Normal condition: Overburden	During earthquake: No overburden	During earthquake: Overburden
Distribution width: b(m)		10.00	10.00	8.77	8.86
Rotational slip	Safety factor	1.76		–	
	Allowable value	1.30		–	
Toe pressure (kN/m^2)	Subgrade reaction	189.86	202.40	359.29	375.96
	Allowable value	450.00		450.00	

Comparison with a case where lightweight treated soil is not used

The width of the caisson in the case with lightweight treated soil is 8.0 m while it is 9.5 m in the case without. The active earth pressure resultant force is 193 kN/m^2 and 280 kN/m^2 under normal conditions, which is a 31% reduction compared to the case without it. During an earthquake, it is 312 kN/m^2 and 416 kN/m^2, which is 25% reduction. As the comparison shows, the use of lightweight treated soil reduces the earth pressure.

- Section diagram

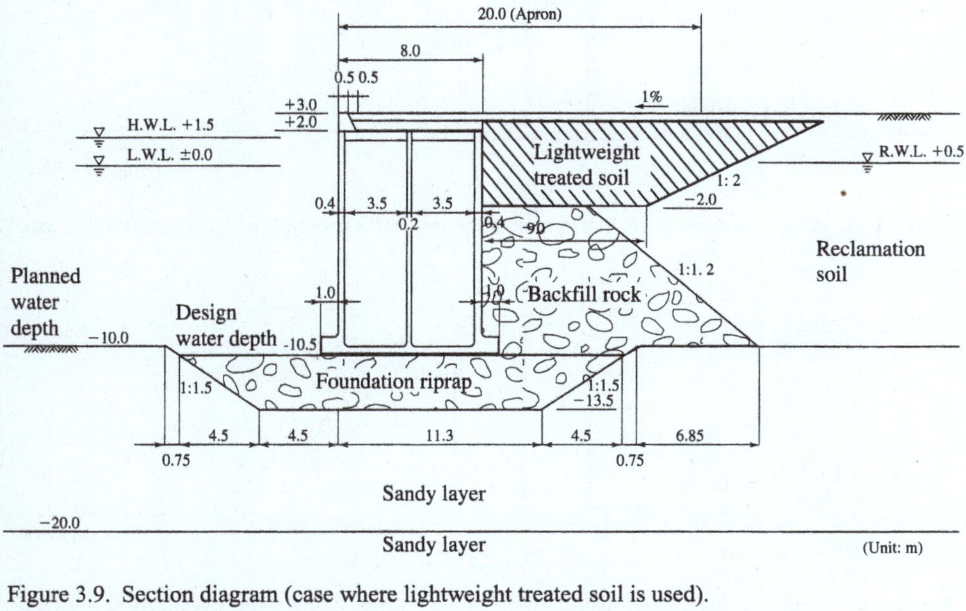

Figure 3.9. Section diagram (case where lightweight treated soil is used).

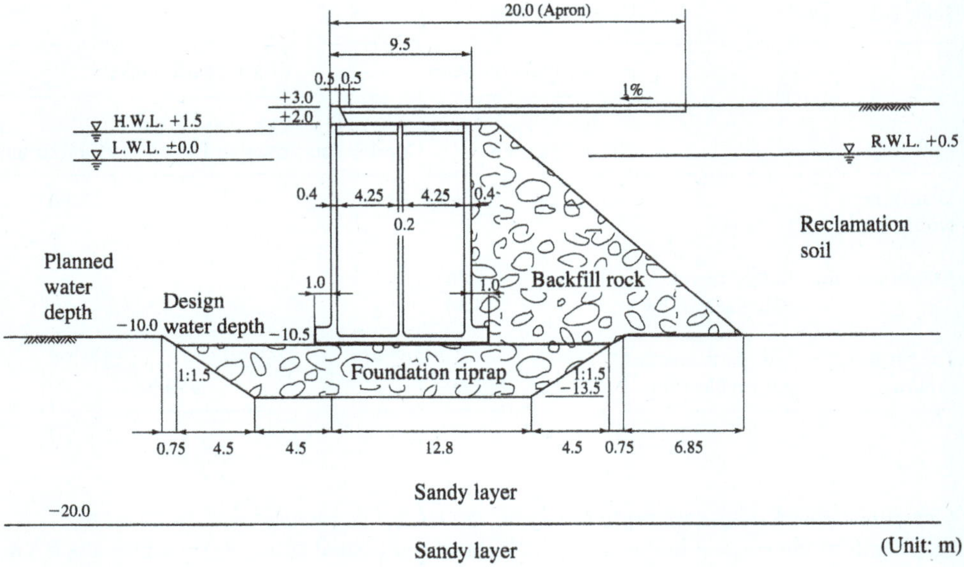

Figure 3.10. Section diagram (case where lightweight treated soil is not used).

- *Comparison of the active earth pressure*
  Normal condition

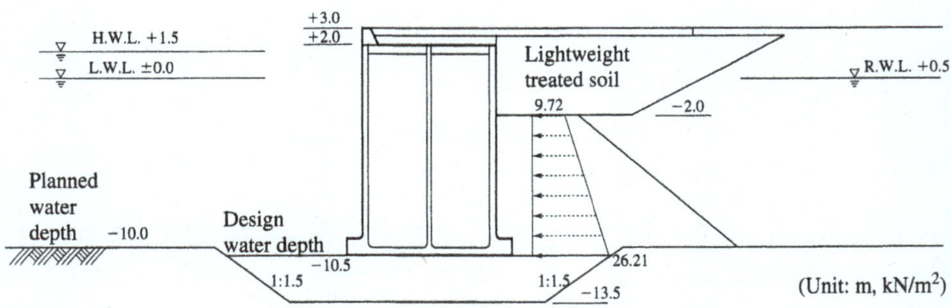

Figure 3.11. Active earth pressure distribution diagram (case where lightweight treated soil is used).

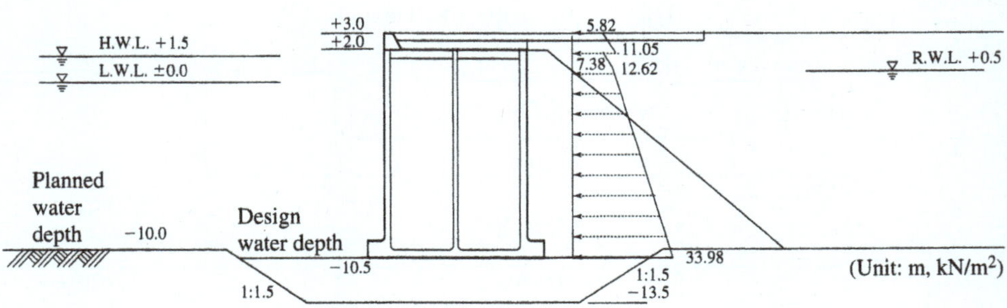

Figure 3.12. Active earth pressure distribution diagram (where lightweight treated soil is not used).

During earthquake

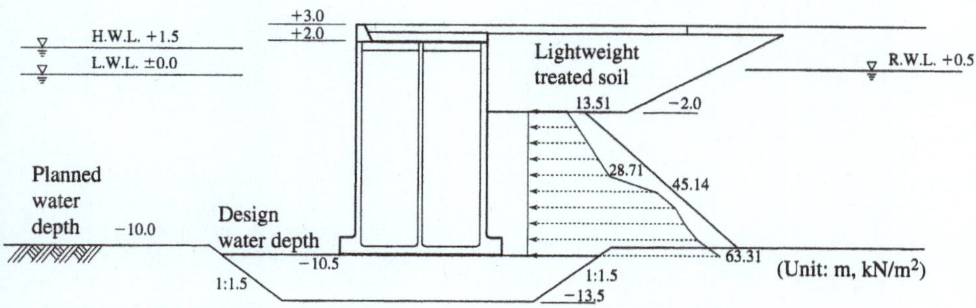

Figure 3.13. Active earth pressure distribution diagram (case where lightweight treated soil is used).

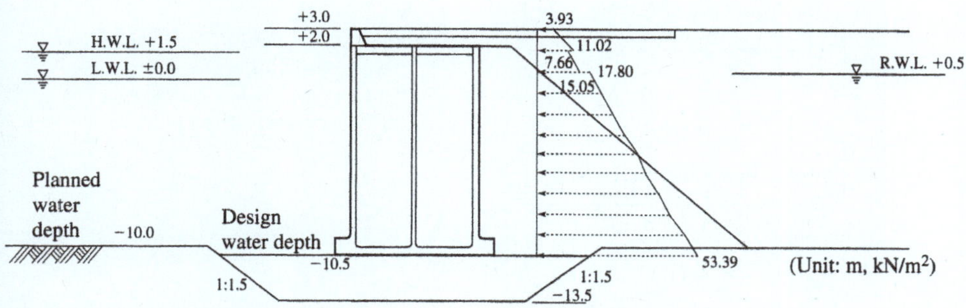

Figure 3.14. Active earth pressure distribution diagram (where lightweight treated soil is not used).

3.4.3 Example of design using lightweight treated soil behind a sheet pile quaywall

Design Procedure

Figure 3.15 is a flow chart of the quay design procedure.

Design conditions

- Design conditions

Tide level	H.W.L. : D.L. +1.5 m L.W.L. : D.L. ±0.0 m R.W.L. : D.L. +1.0 m
Planned water depth	D.L. −7.5 m
Design water depth	D.L. −8.0 m
Crest height	Superstructure crest height: D.L. +3.0 m
Overburden load	Normal: $w = 20\,kN/m^2$ Earthquake: $w' = 10\,kN/m^2$
Design earthquake intensity	Aerial seismic coefficient: 0.15

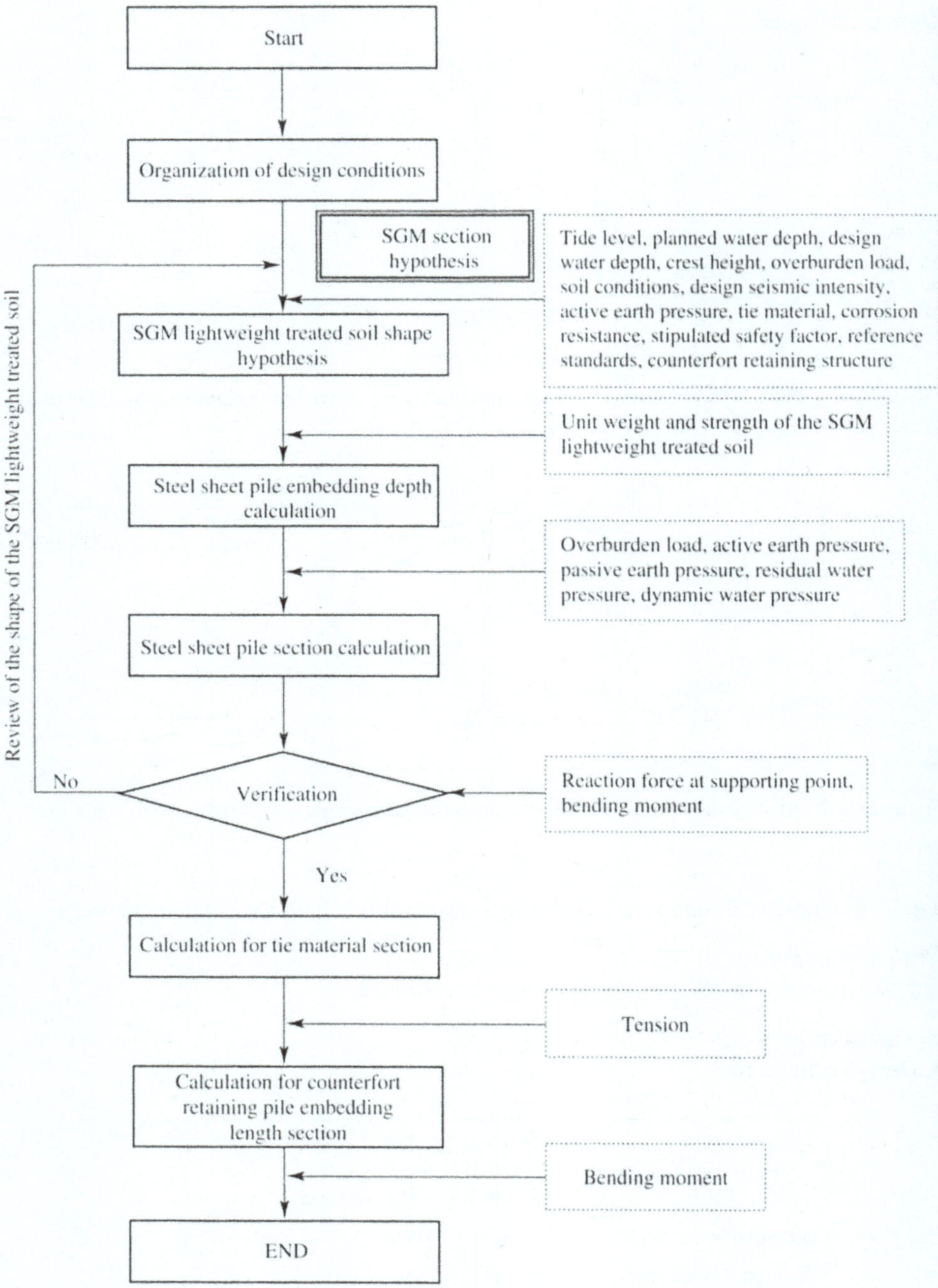

Figure 3.15. Design procedure flow chart.

- Soil conditions

Lightweight treated soil	Weight in the atmosphere:	10.0 kN/m^3
	Weight underwater:	2.0 kN/m^3
	Viscosity:	100.0 kN/m^3
Reclamation soil	Weight in the atmosphere:	18.0 kN/m^3
	Weight underwater:	10.0 kN/m^3
	Internal friction angle:	0.52 rad (30°)
Existing ground	Top layer (D.L. −4.0 m to D.L. −20 m)	
	Weight in the atmosphere:	18.0 kN/m^3
	Weight underwater:	10.0 kN/m^3
	Internal friction angle:	0.52 rad (30°)
	Bottom layer (D.L. −20 m and deeper)	
	Weight in the atmosphere:	18.0 kN/m^3
	Weight underwater:	10.0 kN/m^3
	Internal friction angle:	0.61 rad (35°)
Coefficient of friction of lightweight treated soil and sand: 0.60		

- Earth pressure calculation conditions (active earth pressure)

Normal condition	Earth pressure by slice method
During earthquake	Earth pressure by slice method

- Allowable safety factor

Study item	Normal condition	During earthquake
Sheet pile embedding length	1.5	1.2

- Allowable values

Study item	Normal condition	During earthquake
Steel sheet pile (SY295)	180 N/mm^2	270 N/mm^2
Tie rods (NHT690)	176 N/mm^2	264 N/mm^2

Analysis results
Table 3.2 shows the analysis results.

Comparison with a case where lightweight treated soil is not used
The results from the comparison of the steel sheet pile type reveals the type to be S.P-IV type (geometrical moment of inertia: $I = 38{,}600$ cm^4/m, section modulus: $Z = 2{,}270$ cm^3/m) for

the case with lightweight treated soil and to be S.P-VI$_L$ (geometrical moment of inertia: I = 86,000 cm^4/m, section modulus: Z = 3,820 cm^3/m) for the case without. The steel sheet pile embedding length is 1.5 m less than that with unimproved soil. The use of lightweight treated soil is thus highly effective for reducing the earth pressure.

Table 3.2. Steel sheet pile quaywall calculation results.

Category		Improved	
		Normal condition	During earthquake
Steel sheet pile stress (N/mm^2)	Stress	160	175
	Allowable value	180	270
Embedding length	Safety factor	2.57	1.23
	Allowable value	1.50	1.20
Tie rod tension (N/mm^2)	Tension	173	154
	Allowable value	176	264
Wale stress level (N/mm^2)	Stress	118	105
	Allowable value	140	210
Counterfort steel sheet pile stress level (N/mm^2)	Stress	122	106
	Allowable value	180	270

• Section diagram

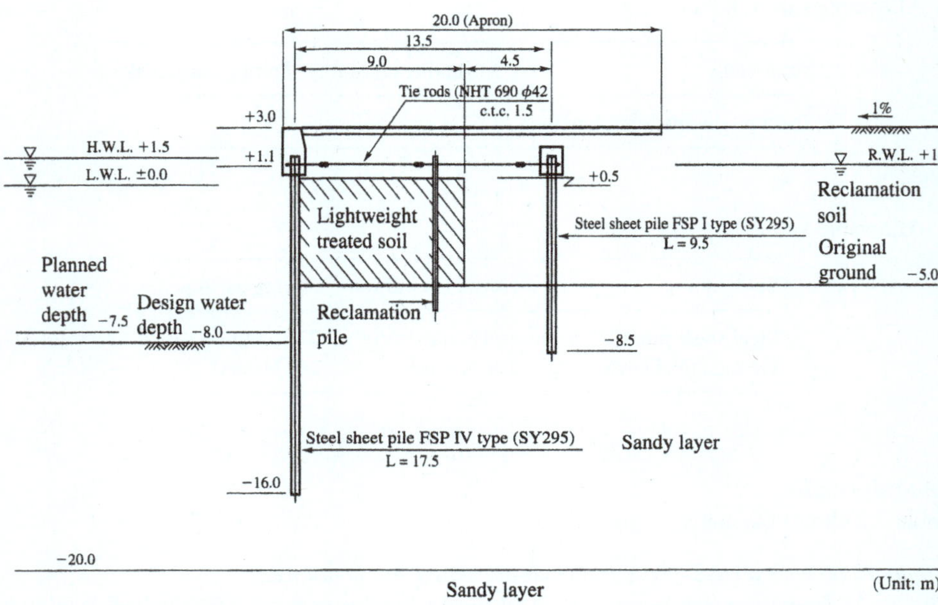

Figure 3.16. Section diagram (case where lightweight treated soil is used).

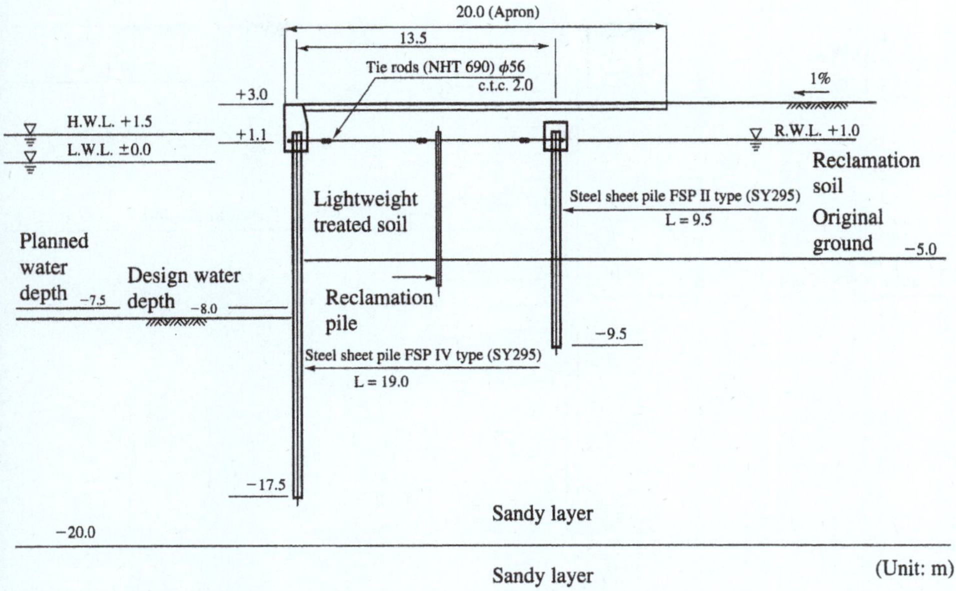

Figure 3.17. Section diagram (case where lightweight treated soil is not used).

- *Comparison of the active earth pressure*
  Normal condition

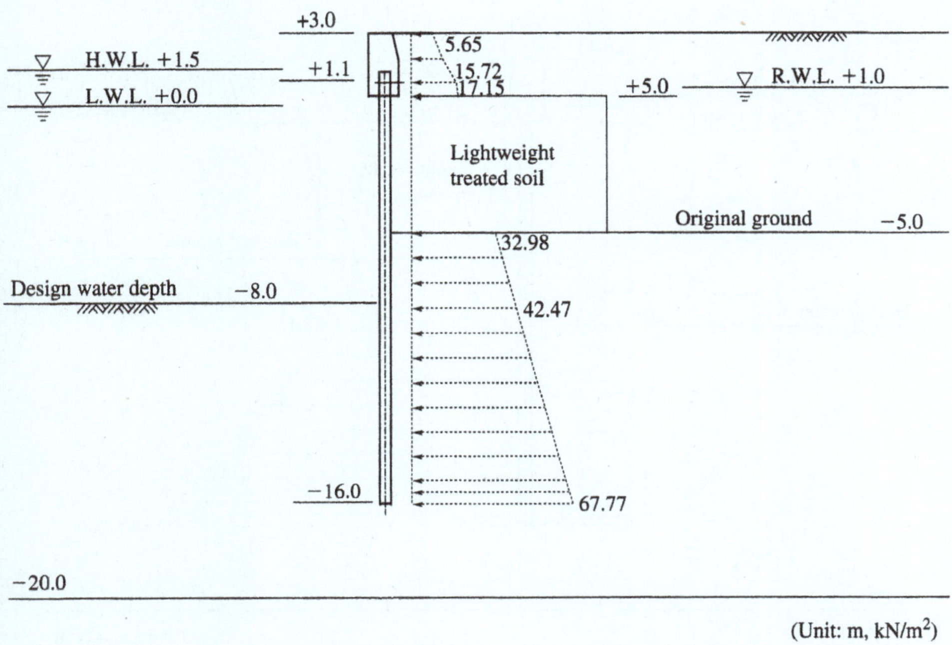

Figure 3.18. Active earth pressure distribution diagram (case where lightweight treated soil is used).

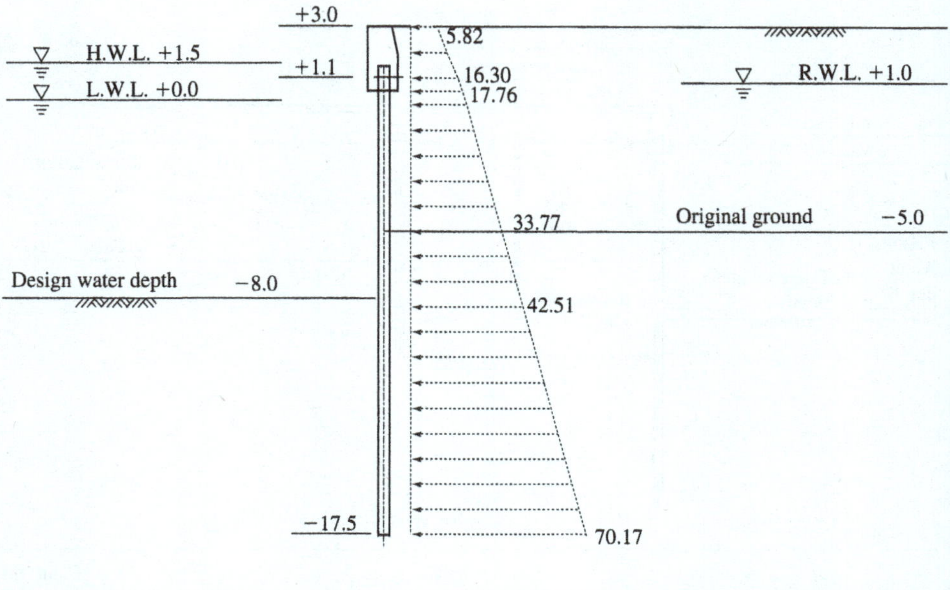

(Unit: m, kN/m²)

Figure 3.19. Active earth pressure distribution diagram (case where lightweight treated soil is not used).

During earthquake

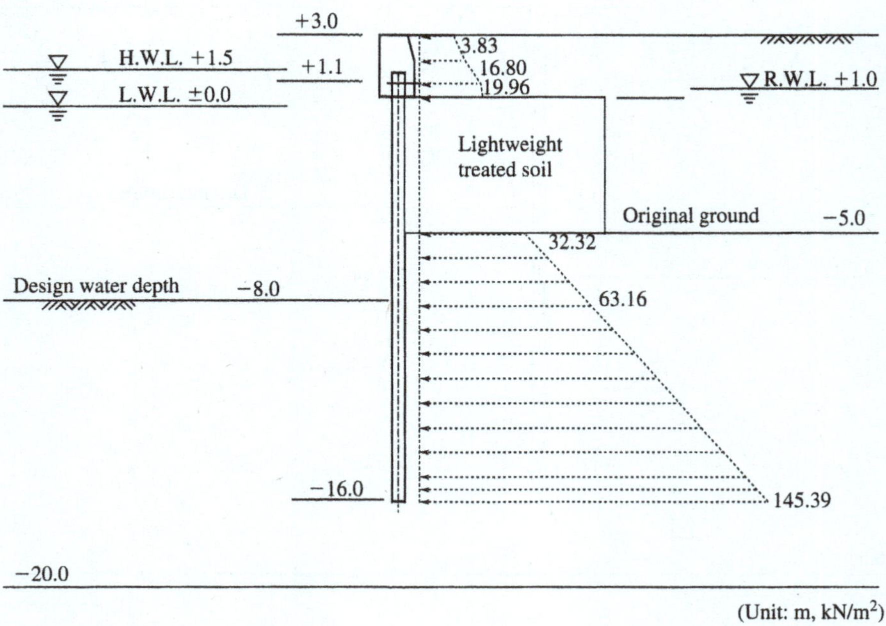

(Unit: m, kN/m²)

Figure 3.20. Active earth pressure distribution diagram (case where lightweight treated soil is used).

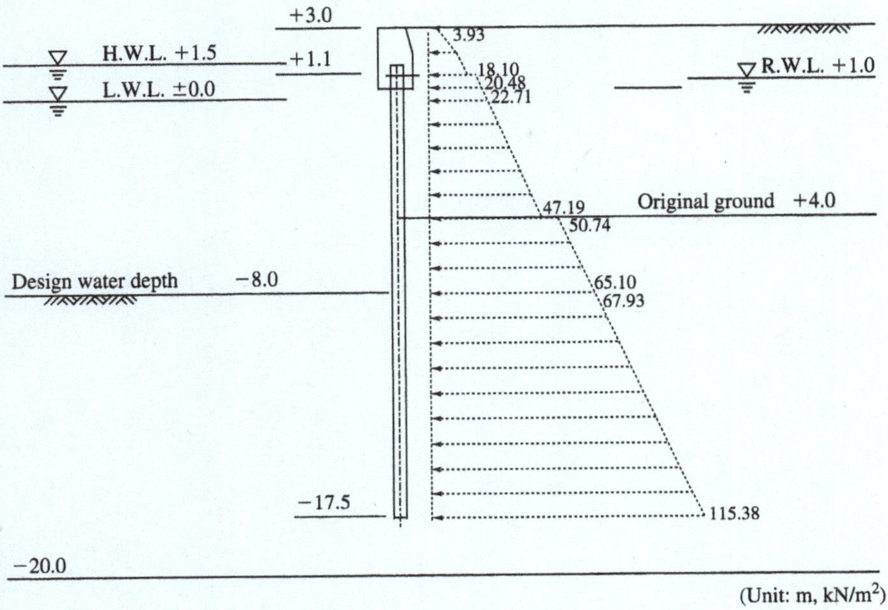

Figure 3.21. Active earth pressure distribution diagram (case where lightweight treated soil is not used).

Chapter 4

Mix Proportion Design

4.1 OUTLINE

Designing the mix proportion of lightweight treated soil is to be performed with mix proportion design values taking density, strength, and flowability as standards.

The density and strength of lightweight treated soil placed underwater are influenced by the water pressure and its movement and segregation properties underwater. Therefore, the mix proportion design of lightweight treated soil should be determined in order to obtain the target strength and density for field application considering the effects of underwater placing.

For example, when air foam treated soil is placed underwater, the water pressure compresses the foam in the treated soil, increasing both the density and strength compared to the ones in the atmosphere. To prevent this, when designing the mix proportion of air foam treated soil, it is necessary to set the mix proportion values considering the increase in the density and strength.

When air foam treated soil is placed in the water, it is defoamed and the water dilutes the soil. Therefore, when designing the mix proportion of air foam treated soil, it is necessary to set mix proportion design values that provide the soil with the viscosity and flowability needed during placing.

When beads treated soil is placed at a depth greater than 10 m, it shrinks due to the low strength EPS beads, affecting the density and strength of the treated soil. This problem can be overcome by using stiffer, stronger EPS according to the water depth.

Beads treated soil floats to the surface as a result of separation of the EPS during underwater placing. Therefore, it is necessary to clarify the quantity of separation of EPS by preliminary tests in order to increase the quantity of EPS to achieve the target density.

4.2 MIX PROPORTION DESIGN OF AIR FOAM TREATED SOIL

4.2.1 Mix proportion design process

The mix proportion of air foam treated soil is designed according to the following procedure. An alternate procedure is necessary under special conditions such as when using some particular source soils or specialized construction methods.

Confirmation of design conditions
To design the mix proportion of air foam treated soil, usage, shape, load conditions, environmental conditions, water depth, execution conditions, and other design conditions are to be established.

Design reference values

Design reference values such as strength, density, flowability, etc. that satisfy the functions of the structure are to be determined considering the long-term material properties of the air foam treated soil.

Mix proportion target values

Mix proportion target values for air foam treated soil are to be determined. The mix proportion target values are the target values of laboratory mix proportion testing intended to correct the design reference values to satisfy the design conditions.

The mix proportion target strength of air foam treated soil is a value obtained as the product of the design strength and an overdesign factor (α). The overdesign factor (α) is represented by the ratio of laboratory mix proportion test strength to design strength. Generally, the overdesign factor $\alpha = 2$ when the treated soil is placed in the atmosphere and $\alpha = 3$ when it is placed underwater. However, even for underwater placing, $\alpha = 2$ can be adopted in a case where the strength in the underwater separation resistance test results fully satisfies the design reference value.

The mix proportion target density is the value obtained by accounting for the rise in the density under the effects of the execution by reducing the design density by the density increase ($\Delta\rho$). The rise in the density is caused by defoaming during the placing, initial shrinkage, the absorption of water after hardening, and so on. In Japan, the increase in the density is assumed to be $\Delta\rho = 0.1$ g/cm^3 as the standard.

The mix proportion target flowability value is set accounting for the field placing conditions.

Determination of the basic mix proportions

Mix proportions are to be selected with reference to soil test results and the results of previous mix proportions, in order to satisfy the mix proportion target values. Taking several mix proportions selected as the basic mix proportions, the quantities of each material in the mix are determined by calculations.

For the basic mix proportions of air foam treated soil, the water content of a source soil ranges from 1.5 to 3.0 W_L and a stabilizing agent quantity ranges from 50 to 200 kg/m^3 in general.

Mix proportion test

The mix proportion test is carried out to confirm that the density of the adjusted slurry conforms with the calculated value. In the mix proportion test for air foam treated soil, density, flowability, strength, and material separation should be confirmed. When it is assumed that the test will be affected by air temperature during placing, the mix proportion test is carried out using a method that accounts for curing conditions during hardening.

Determining the modified mix proportion

Based on the results of the mix proportion test, the mix proportion is modified in order to obtain the specified physical properties. The strength is modified primarily by varying the quantity of stabilizing agent. The density is modified by adjusting the quantity of air foam. For flowability, the water content of the adjusted slurry is controlled for modification.

After these modifications, the modified mix proportion is determined to meet the required physical properties.

Determining the field mix proportion

The field mix proportion is determined by modifying the quantity of air foam based on the results of measuring the defoaming rate during mix proportion tests in in-situ placing and by increasing the quantity of air foam based on the placing depth.

The quantity of air foam in the field mix proportion is increased in cases where it is anticipated that the air foam volume will shrink due to the water pressure. This increase is calculated based on the following concept.

If the air foam treated soil is compressed prior to hardening, only the air in the air foam shrinks, lowering the volume. Therefore, the increase in the density may be dealt with by adding the quantity of compressed air foam equal to the volume reduction. The increase in the quantity of air foam is obtained by adding this to the quantity of air foam under atmospheric pressure.

Generally, using Boyle's Law, the volume V_h of the air foam at water depth $h(m)$ can be represented by the following equation, where the volume of the air at atmospheric pressure is represented by V_0.

$$V_h = \{1/(1 + h/10)\} \cdot V_0$$

4.2.2 Mix proportion testing process

A mix proportion test of air foam treated soil is carried out using the materials used in the actual field in order to obtain the proportions of stabilizing agent, water, air foam, and other additives that satisfy the design.

The mix proportion test is performed by the following procedure in order to obtain the target physical properties.

Property test of the source soil

The physical properties of the source soil are clarified by carrying out the tests on density, particle size distribution, liquid and plastic limit, water content, wet density, *pH*, organic content, and ignition loss.

Determining the water content of the adjusted slurry

Three kinds of adjusted slurry are decided before the mix proportion test in order to obtain the target flowability. The adjusted slurry at this time is set with 2.5 times the liquid limit as the criterion.

Determining the quantity of stabilizing agent

Three kinds of stabilizing agent mix proportions are decided in order to achieve the target strength.

Mix proportion calculation

The quantity of air foam is calculated in order to achieve the target density.

Foaming agent evaluation test (selection of the foaming agent)

Air foam treated soil is prepared by varying the type of foaming agent to confirm the degree of defoaming during and after mixing. Based on the results of this test, a foaming agent that minimizes the effects of agitation is selected.

These trends vary according to the conditions of the materials and construction equipment used in field work, so the foaming agent should be confirmed by trial mixing.

Physical properties test

The foaming agent selected based on the results of the foaming agent evaluation test is used to prepare the treated soil by the procedure shown in Figure 4.1. The physical properties test is carried out by varying the water content of the adjusted slurry and the quantity of stabilizing agent added in order to confirm their effects on the strength development or flowability of the treated soil and other fundamental physical properties.

To evaluate the effects of the water content of the adjusted slurry and the quantity of stabilizing agent, tests are carried out under three sets of conditions to obtain the mix proportion that provides the target density. The underwater separation resistance test is then carried out to confirm the separation resistance, *pH*, *SS*, and other water qualities.

The following is the mix proportion test procedure:

(i) *Density.* The mass capacity of the specimens is measured with a quantity measuring vessel. For example, air foam treated soil is placed in a 1-liter vessel to measure its mass.

(ii) *Flowability.* This is based on the cylinder method (diameter: 80 mm, height: 80 mm) in JHS A 313 Air Mortar and Air Milk Test Method, with the load technique of the same

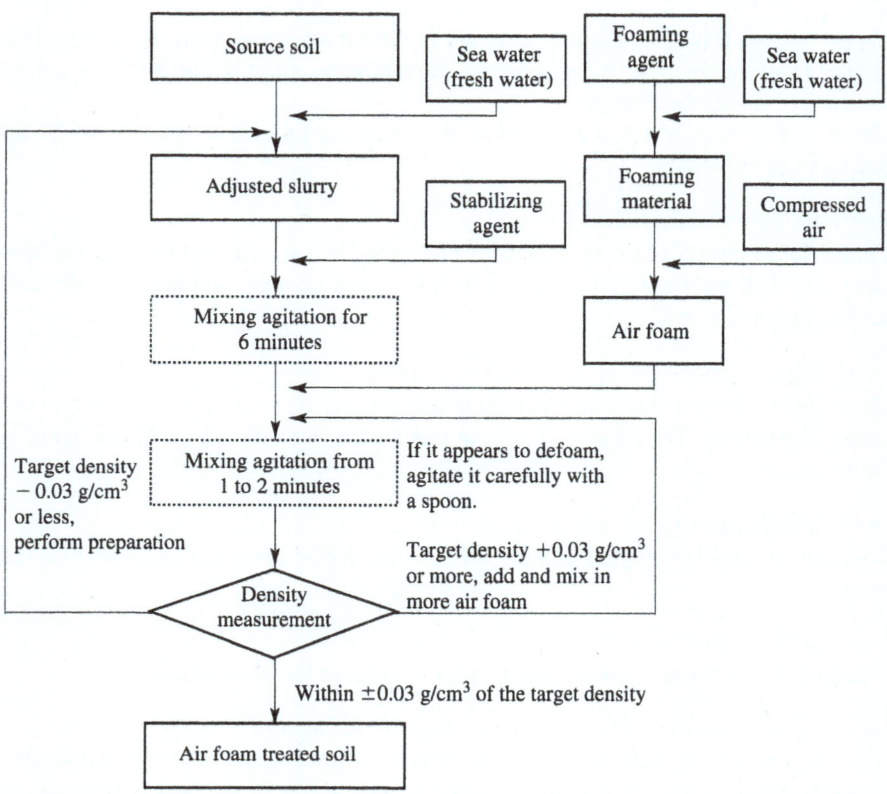

Figure 4.1. Preparation of air foam treated soil for mix proportion testing.

test method, or with the JSCE-F 521-1994 Test Method for Injection Mortar of Prepacked Concrete (method using a P-funnel). When the test is carried out underwater, using the cylinder system, a cylinder is placed in the center of the bottom of a 16-cm deep acrylic water tank. The specimen is poured into it, the tank is then filled with artificial sea water to a depth of 16 cm, and the flow value is measured in the same way as in the atmosphere.

(iii) *Unconfined compressive strength.* This is carried out based on the JIS A 1216 Unconfined Compressive Test Method for Soil. After the specimen packed in its mold is sealed with wrap, it is cured in an air controlled tank maintained at a temperature of 20 ± 3°C and a humidity close to 100%. The standard aging periods are 7 days and 28 days.

Pressurized curing to simulate water pressure is performed separately by an additional method.

(iv) *Wet density.* This is measured in conformity with the JGS T 191 Wet Density Test Method of Soil.

(v) *Confirming the underwater separation resistance and contamination in water.* To provide a simple laboratory test method to determine the suitability of underwater placing, the SGM Lightweight Soil Association has proposed the underwater separation resistance test that can be used to examine the separation resistance properties of lightweight treated soil according to its flowability and placing rate.

4.2.3 Sample mix proportion test

Materials

- Source soil: soil excavated at the site;
- Water (diluted water, water content adjusting water): sea water obtained near the site;
- Stabilizing agent: blast-furnace slag cement type B;
- Foaming agent: Surface-active agent, protein type.

Target physical properties

Table 4.1. Target physical values of air foam treated soil.

Category	Unconfined compressive strength	Wet density	Flow value in the atmosphere
Design values	200 kN/m^2 or more (aging for 28 days)	1.20 t/m^3 (after hardening)	–
Laboratory mix proportion target values	600 kN/m^2 or more (aging for 28 days)	1.20 t/m^3 (after hardening)	18 cm [18 ± 2 cm]

Test results

- Results of physical and chemical property tests for the source soil.
  The test results are shown in Table 4.2.

Table 4.2. Results of the physical and chemical property tests for the source soil.

	Test item	Unit	
Density/water content	Wet density ρ_t	t/m^3	1.349
	Soil particle density ρ_s	t/m^3	2.655
	Natural water content ratio W_n	%	140.8
Particle size distribution	Gravel (2 to 75 mm)	%	1
	Sand (75 μm to 2 mm)	%	12
	Silt (5 to 75 μm)	%	25
	Clay (less than 5 μm)	%	62
Consistency properties	Liquid limit W_L	%	91.2
	Plastic limit W_P	%	33.7
	Plastic index I_P		57.5
Classification	Code	–	(CH)
Chemical properties	Ignition loss test	%	8.7
	pH test		7.9
	Organic content	%	3.50

Table 4.3. Adjusted slurry density and flow value test results.

Adjusted slurry density (g/cm^3)	Flow value in atmosphere (cm)
1.311	24.2
1.269	32.3
1.251	37.5
1.237	42.5
1.207	47.0

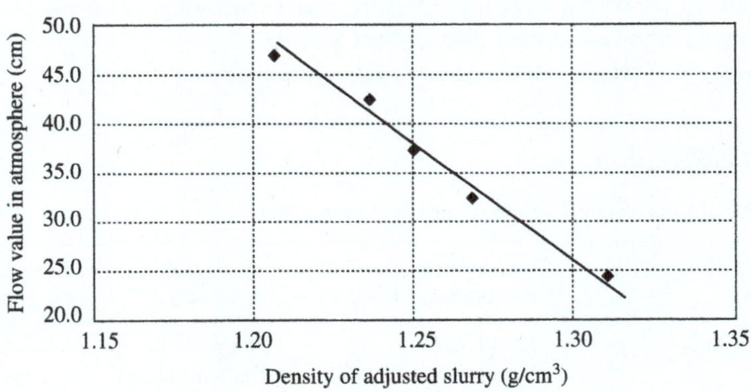

Figure 4.2. Adjusted slurry density and flow value test results.

- Results of tests for the adjusted slurry.
  The results of the tests on the density and flow value of the adjusted slurry are shown in Table 4.3 and in Figure 4.2, and the adjusted slurry density and water content that have been determined are shown in Table 4.4.

Table 4.4. Set density and water content of adjusted slurry.

Flow value (cm)	Wet density (g/cm^3)	Water content (%)	$\times W_L$
35 (33–37)	1.26	223	2.44
40 (38–42)	1.24	247	2.70
45 (43–47)	1.22	275	3.03

Table 4.5. Set density and water content of adjusted slurry.

Protein type				Surface-active agent type			
Air foam quantity (%)		Wet density (g/cm^3)	Flow value (cm)	Air foam quantity (%)		Wet density (g/cm^3)	Flow value (cm)
(Fixed	100	1.118		(Fixed	100	1.201	
quantity)	110	1.110		quantity)	160	1.157	
	120	1.081			200	1.112	
	140	1.049	18.8		240	1.073	20.3

- *Foaming agent stability test results*
  Test conditions:
  - Water content of adjusted slurry: $2.70 \cdot W_L$
  - Foaming agents: protein type, surface-active agent type
  - Target density of air foam treated soil: $\rho_t = 1.10\,g/cm^3$
  - Quantity of foaming agent added: fixed quantity, increased in 3 stages, total of 4 stages
  - Measurement: after 2 minutes of hand mixing

As the test results in Table 4.5 show, the air foam produced by the surface-active agent type could not obtain the predetermined target density even if the air foam was increased 100% or more. However, the 10 to 20% increase in air foam provided by the protein type could achieve the target density. Therefore, the protein type is used as the foaming agent.

- *Test for the quantity of stabilizing agent added*
  Test conditions:
  - Water content of adjusted slurry: 2.44, 2.70 and $3.03 \cdot W_L$
  - Foaming agent: protein type
  - Stabilizing agent: blast-furnace slag cement type B
  - Target flow value in the atmosphere (range): 18 cm (18 cm $\pm$ 2 cm)
  - Target density of the air foam treated soil: $\rho_t = 1.10\,g/cm^3$
  - Target strength of laboratory mix proportion of air foam treated soil (unconfined compressive strength): 600 kN/m^2 as minimum
  - Quantity of cement added: 60 kg, 90 kg and 120 kg

The test for the quantity of stabilizing agent added used a total of nine basic mix proportions set by combining three adjusted slurry water contents with three quantities of cement added. The following is an example of the basic mix proportion calculation method.

- *Calculating the basic mix proportions*

  The basic mix proportions determined by the following conditions are calculated:

  Water content of the adjusted slurry: $W_t = 222.5\%\ (2.44 \cdot W_L)$
  Unit weight of the stabilizing agent: $m_c = 90\,\text{kg/m}^3$
  The following are the target densities and the properties of the materials.
  Target density: $\rho = 1.10\,\text{g/cm}^3$
  Soil particle density: $\rho_s = 2.655\,\text{g/cm}^3$
  (Sea) water density: $\rho_w = 1.03\,\text{g/cm}^3$
  Stabilizing agent density: $\rho_c = 3.05\,\text{g/cm}^3$

Mass related equation $m_s + m_w + m_c + n_1 \cdot m_A = 100 \cdot \rho$
Volume related equation $m_s/\rho_s + m_w/\rho_w + m_c/\rho_c + n_1 \cdot n_2 \cdot m_A = 1000$

m_s: unit mass of dry soil (kg/m^3)
m_w: unit mass of (sea) water (kg/m^3); $m_w = W_t \cdot m_s/100$
m_A: unit mass of foaming agent (kg/m^3)
n_1: dilution multiplying rate $= 10$
n_2: foaming multiplying rate $= 25$

The density of the adjusted slurry is represented as follows.

$$\rho_t = \frac{m_s + m_w}{m_s/\rho_s + m_w + \rho_w} = \frac{3.225\, m_s}{m_s/2.655 + 2.160 \cdot m_s} = 1.27\,\text{g/cm}^3$$

The unit mass of foaming agent, dry soil, and (sea) water are respectively calculated as:

$$m_A = \frac{1000 \cdot (\rho_t - \rho) + m_c \cdot (1 - \rho_t/\rho_c)}{n_1 \cdot (n_2 \cdot \rho_t - 1)} = 0.724\,\text{kg/m}^3$$

$$m_s = \frac{1000 \cdot \rho - n_1 \cdot m_A - m_C}{3.225} = 311\,\text{kg/cm}^3$$

$$m_w = 1000 \cdot \rho - n_1 \cdot m_A - m_C - m_s = 1100 - 7 - 90 - 311 = 692\,\text{kg/cm}^3$$

If the volume is calculated subsequently,

Air foam	$n_1 \cdot m_A = 7$	$n_1 \cdot n_2 \cdot m_A = 181$
Dry soil	$m_s = 311$	$m_s/\rho_s = 117$
Water (sea water)	$m_w = 692$	$m_w/\rho_w = 672$
Stabilizing agent	$m_c = 90$	$m_c/\rho_c = 30$
Total	1,100 kg	1,000 *l*

The volume of foaming agent is to be increased considering the defoaming rate based on trial mixing. The basic mix proportion is shown in Table 4.6 for the conditions that the water content of the adjusted slurry is 222.5% and the unit mass of the stabilizing agent added is 90 kg/m^3.

Table 4.6. Basic mix proportion.

	Material				
Category	Dry soil	Sea water	Stabilizing agent	Air foam	Total
Unit mass (kg/m^3)	311	692	90	7.2	1,100
Unit volume (l/m^3)	117	672	30	181	1.000

Table 4.7. Results of testing to determine the quantity of stabilizing agent added.

Adjusted slurry			Physical property values of air foam treated soil		
Water content $\times W_L$	Wet density (g/cm^3)	Quantity of cement C (kg/m^3)	Flow value in atmosphere (cm)	Wet density ρ_{28} (g/cm^3)	Unconfined compressive strength q_{u28} (kN/m^2)
2.44 (223%)	1.260	60	14.7	1.108	195
		90	14.0	1.092	544
		120	14.0	1.089	1,027
2.70 (247%)	1.240	60	16.9	1.078	150
		90	17.5	1.080	564
		120	16.8	1.079	805
3.03 (275%)	1.221	60	22.9	1.082	171
		90	22.3	1.076	510
		120	21.9	1.077	626

Test results

The results of calculating the nine basic mix proportions and their tests are shown in Table 4.7 and in Figures 4.3 and 4.4.

Studying the test results

Judging from the relationship between the adjusted slurry water content and the flow value in the atmosphere shown in Figure 4.3, at the same adjusted slurry water content, the flow value remains almost constant when the quantity of cement added is increased. From the comparison at a constant quantity of cement, it is found that as the adjusted slurry water content increases, the flow value also increases.

The adjusted slurry water content ratio that satisfies the target flow value in the atmosphere (18 cm $\pm$ 2 cm) was found to range from $2.72 \cdot W_L$ to $2.92 \cdot W_L$ regardless of the quantity of cement. The average adjusted slurry water content is determined at $2.8 \cdot W_L$.

For the cases of 2.70 W_L and 3.03 W_L in Figure 4.4, taking the values at the unconfined compressive strength, q_u, of larger than 600 kN/m^2, the adding quantity of cement can be determined as 94.5 kg/m^3 and 113.3 kg/m^3, respectively.

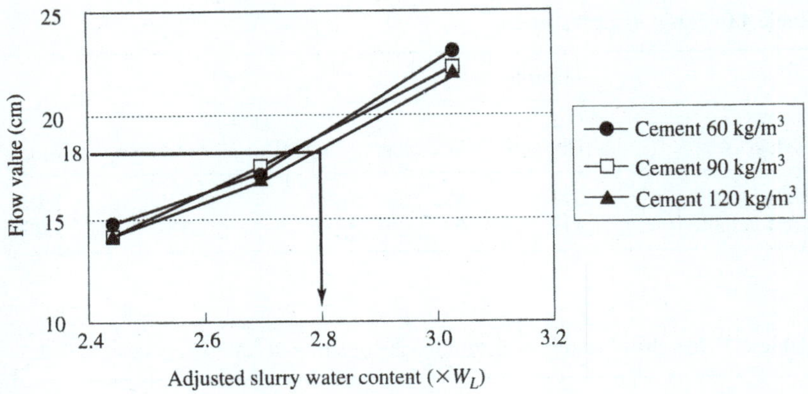

Figure 4.3. Adjusted slurry water content–flow value in atmosphere relationship.

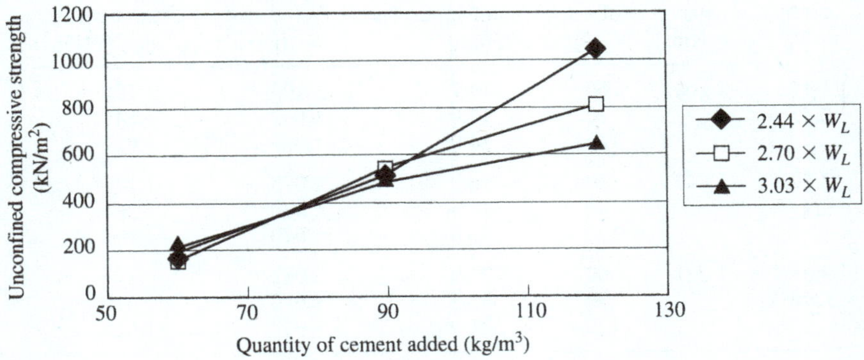

Figure 4.4. Quantity of cement added–unconfined compressive strength relationship.

Underwater separation resistance testing

- Mix proportion conditions

Adjusted slurry water content (× W_L)	Quantity of cement (kg/m^3)
2.70	90
3.03	90
3.03	120

- Measurement

  Density variation: The difference between the wet density values after mixing and after underwater placing (aged for 7 days) is obtained by the underwater separation resistance test to confirm if it provides underwater separation resistance. The index that can ensure the predetermined material properties is the density difference less than 0.05 g/cm^3 of the lightweight treated soil caused by underwater placing.

Table 4.8. Underwater separation resistance test results.

Adjusted slurry water content ratio ($\times W_L$)	Cement quantity (kg/m^3)	Density fluctuation $\Delta\rho'$ (g/cm^3)	pH	SS	Strength decline rate (q'_{u7}/q_{u7})	Predicted unconfined compressive strength q_{u28} kN/m^2
2.70	90	0.019	9.1	40	0.414	234
3.03	90	0.014	9.4	80	0.480	245
3.03	120	0.026	9.7	70	0.798	500

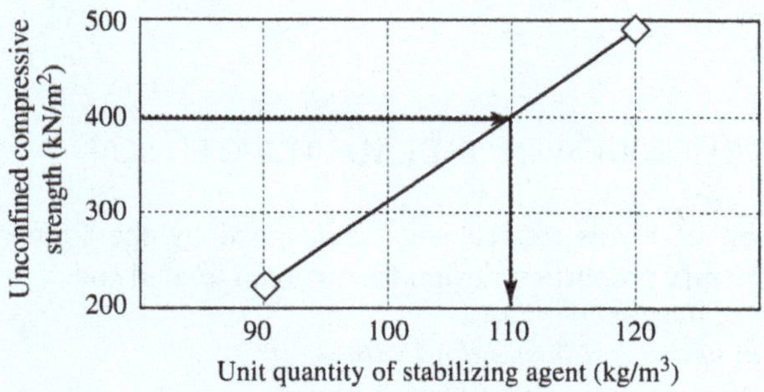

Figure 4.5. Unit quantity of stabilizing agent–unconfined compressive strength relationship.

Water quality analysis: pH of 10.5 or less and SS of 100 mg/l also indicate that lightweight treated soil placed underwater satisfies its predetermined material properties.
Unconfined compressive strength variation: The unconfined compressive strength (aged for 28 days) is obtained by evaluating the average unconfined compressive strength obtained by the underwater separation resistance test and rate of increase of strength after placing in the atmosphere.

- Test results

The results of density variation and water quality analysis from the underwater separation resistance test in Table 4.8 are found to be all within permissible ranges. The standard mix proportion is assumed to be the unit quantity of stabilizing agent that provides unconfined compressive strength of 400 kN/m^2 (safety factor $\alpha = 2.0$) that satisfies the basic design values, and 110 kg/m^3 of cement is added based on Figure 4.5.

Setting the standard mix proportion

Table 4.9. Properties of adjusted slurry.

Water content (with reference to the liquid limit)	Water content W (%)	Saturated density ρ (g/cm^3)
$2.8 \times W_L$	255	1.234

Table 4.10. Mix proportion.

	Water content ratio, adjusted slurry				Lightweight material			
	Dry soil m_s	Water m_w	Total	Stabilizing agent m_c	Foaming agent m_A	Dilution water m_W	Total [air foam]	Total
Mass (kg)	(276)	(705)	981	110	(0.485)	(8.165)	8.650	1,100
Volume (m^3)	0.094	0.622		0.028			0.257	1.000

4.3 MIX PROPORTION DESIGN OF BEADS TREATED SOIL

The mix proportion of beads treated soil is designed by the following procedure in accordance with the mix proportion design for air foam treated soil:

(i) *Confirmation of design conditions*
(ii) *Determination of the design standard values*
(iii) *Determination of mix proportion target values*

It has been confirmed that the target density of the mix proportion for beads treated soil is increased slightly by contraction caused by separation of the EPS beads and pump pressure feeding during placing, but there is almost no change of the density caused by water pressure during placing so that the value is considered to be equal to the design density.

Determination of the basic mix proportions

(iv) The basic mix proportions of beads treated soil are to be selected to obtain the target mix proportions with reference to soil test results and to the results of previous mix proportion tests.

- Example of mix proportion calculation:

  The mix proportion is calculated under the following conditions;
  Target density: $\rho = 1.20\ \text{g/cm}^3$
  Adjusted slurry water content: $W_t = 150\%$
  Unit stabilizing agent content: $m_c = 100\ \text{kg/m}^3$
  Soil particle density: $\rho_s = 2.60\ \text{g/cm}^3$
  (Sea) water density: $\rho_w = 1.03\ \text{g/cm}^3$
  Stabilizing agent density: $\rho_c = 3.05\ \text{g/cm}^3$
  EPS bead density (foaming multiplier: 30×): $\rho_b = 0.0533\ \text{g/cm}^3$

  Mass related equation $m_s + m_w + m_c + m_b = 1000 \cdot \rho$
  Volume related equation $m_s/\rho_s + m_w/\rho_w + m_c/\rho_c + \rho_b \cdot m_b = 1000$

  m_s: unit weight of dry soil (kg/m^3)
  m_w: unit weight of (sea) water (kg/m^3) $m_w = W_t \cdot m_s/100$
  m_b: unit weight of EPS beads (kg/m^3)

Table 4.11. Basic mix proportions.

Category	Materials				
	Dry soil	Sea water	Stabilizing agent	Foam beads	Total
Weight (kg/m^3)	437.3	655	100	7.7	1,100
Volume (l/m^3)	168	655	33	144	1,000

Note: Adjusted slurry = Dry soil + sea water.

The density of the adjusted slurry is represented below.

$$\rho_t = \frac{m_s + m_w}{m_s/\rho_s + m_w/\rho_w} = \frac{2.5 \cdot m_s}{m_s/2.6 + 1.5 \cdot m_s} = 1.33 \text{ g/cm}^3$$

If these are subsequently calculated:

EPS	$m_b = 7.7$	$m_b/\rho_b = 144$
Dry soil	$m_s = 437$	$m_s/\rho_s = 168$
(Sea) water	$m_w = 655$	$m_w/\rho_w = 655$
Stabilizing agent	$m_c = 100$	$m_c/\rho_c = 33$
Total	1,200 kg	1,000 l

Table 4.11 shows the basic mix proportion design values and mix proportions of beads treated soil.

Mix proportion test

The items tested and the test method used during trial mixing of beads treated soil are based on the mix proportion test method for air foam treated soil except the foaming agent evaluation test.

Determining the modified mix proportion

The density is modified by adjusting the quantity of EPS beads.

Determining the field mix proportion

The field mix proportion is controlled by increasing the quantity of EPS in cases where there is a risk of the volume shrinkage of EPS by the pump feed pressure during construction or the separation of the EPS during underwater placing.

Chapter 5

Application

5.1 OUTLINE

5.1.1 Flow of application

The application of lightweight treated soil consists of the procedures of dredging, transport, slurrying, mixing, placing, and curing (see Fig. 5.1).

It is necessary to select construction machinery appropriate to the conditions and the scale of application in order to maintain the specified quality of the lightweight treated soil throughout all of these steps, and to carry out the application taking account of the separation during the placing.

The following are the major processes in the application of lightweight treated soil:

- *Dredging and transport work.* In the dredging and transport work, the source soil (dredged soil, etc.) for the lightweight treated soil is obtained and is transported to the application site.

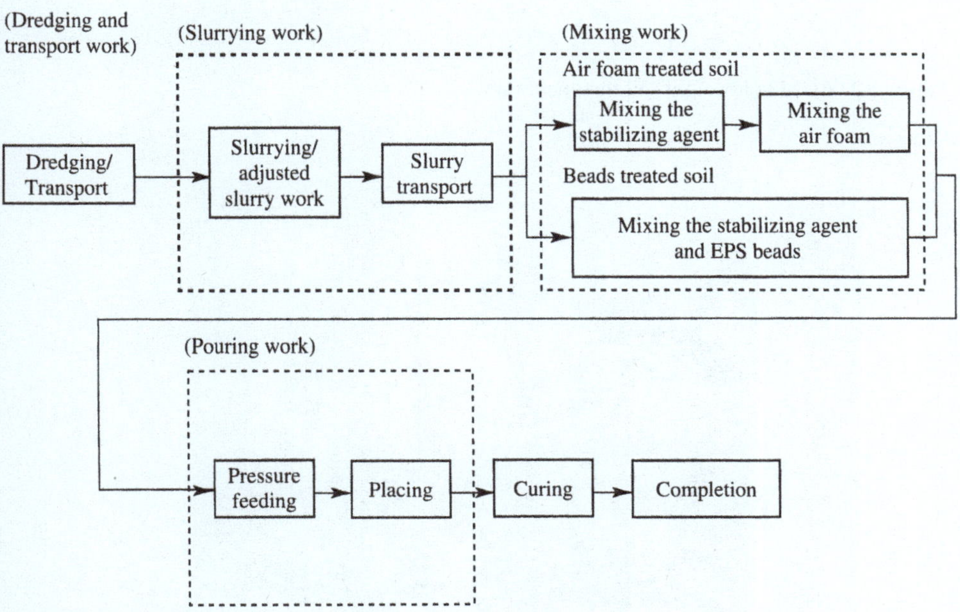

Figure 5.1. Lightweight treated soil application procedure.

- *Slurrying work.* In slurrying work, the source soil from which extraneous materials are removed, is agitated with water admixed to form slurry in order to adjust its water content and density to the specified values.
- *Mixing work.* The lightweight material (air foam or EPS beads) and stabilizing agent are added and mixed with the source soil in this process after its water content and density have been adjusted.
- *Placing work.* Placing is carried out by pumping the mixed material through a tremie pipe, etc. under pressure to the specified location.
- *Curing.* The treated soil after placing is cured for the specified period of time.

5.1.2 Application system

Figure 5.2 shows an example of the application system corresponding to each process explained above, and Photograph 5.1 shows a complete plant. The manufacturing plant shown in Photograph 5.1 corresponds to process 3 in Figure 5.2.

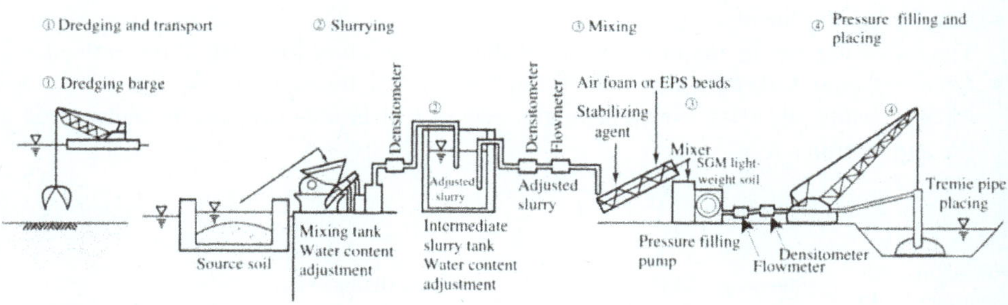

Figure 5.2. SGM lightweight soil treatment flow.

Photograph 5.1. Manufacturing plant.

5.2 APPLICATION AND CONSTRUCTION METHOD

5.2.1 Slurrying

5.2.1.1 *Process of slurrying*

Figure 5.3 shows the flow of the slurrying process.

- *Supply.* The source soil is transported to the site by a soil barge or by a dump truck. The source soil brought to the site is placed in a hopper by a back hoe, then unnecessary extraneous materials are removed by screening using a vibrating sieve, etc.
- *Slurrying.* The source soil screened is thoroughly mixed by a slurrying device, while adjusting the water content to the specified range. The slurrying process continues until the slurry becomes dispersed (about 10 mm size) not to obstruct the later mixing, feeding, and placing steps.
- *Slurry adjustment.* The prepared source soil is placed in the slurry adjustment tank and water is added to increase its water content by between 1.5 and 2.5 times the liquid limit, then it is agitated until its density and water content are uniformly at the specified values. The source soil whose density and water content have been adjusted to the specified values in this way is called adjusted slurry.
- *Slurry storage (Photograph 5.2).* The adjusted slurry is stored in the slurry storage tank. To prevent the soil particles from settling in the bottom of the tank before the material

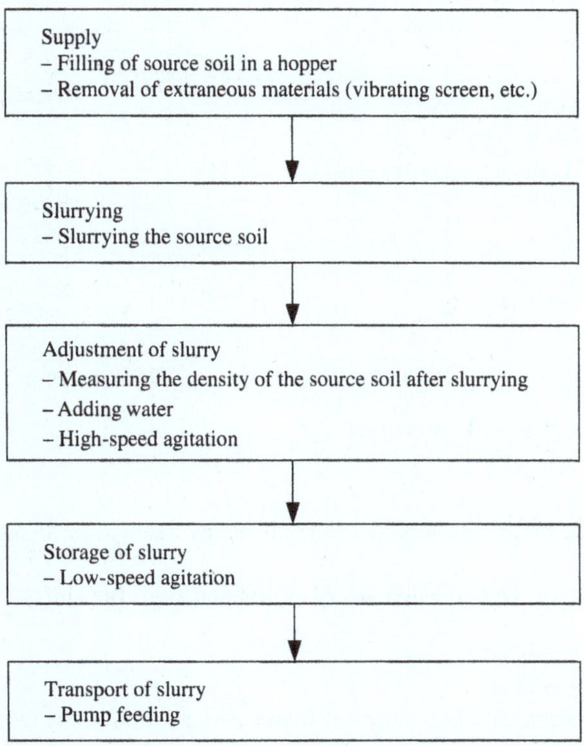

Figure 5.3. Slurrying work procedure.

Photograph 5.2. Example of a slurry storage tank.

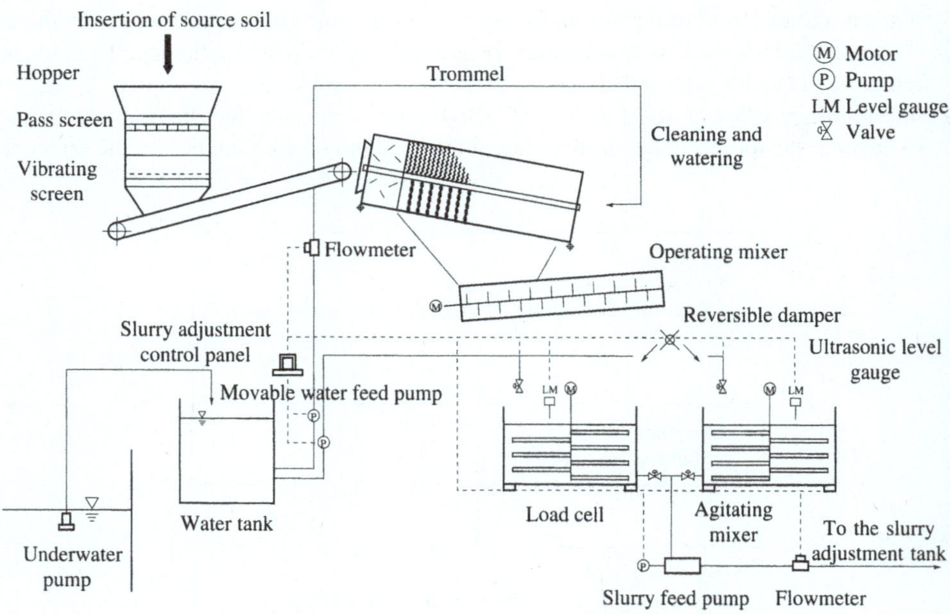

Figure 5.4. Example of a slurrying system.

is supplied to the mixer, the adjusted slurry is gently agitated by an underwater mixer or agitator.

- *Transport of slurry.* The adjusted slurry is pumped from the slurry storage tank to the mixing plant.

5.2.1.2 *Slurrying system*

In the slurrying system, adjusted slurry of source soil is produced by removing wood chips and other unwanted material and concurrently adjusting the water content and density of the source soil to the specified value. Then this material is fed to the slurry adjustment tank.

Photograph 5.3. Slurrying operation.

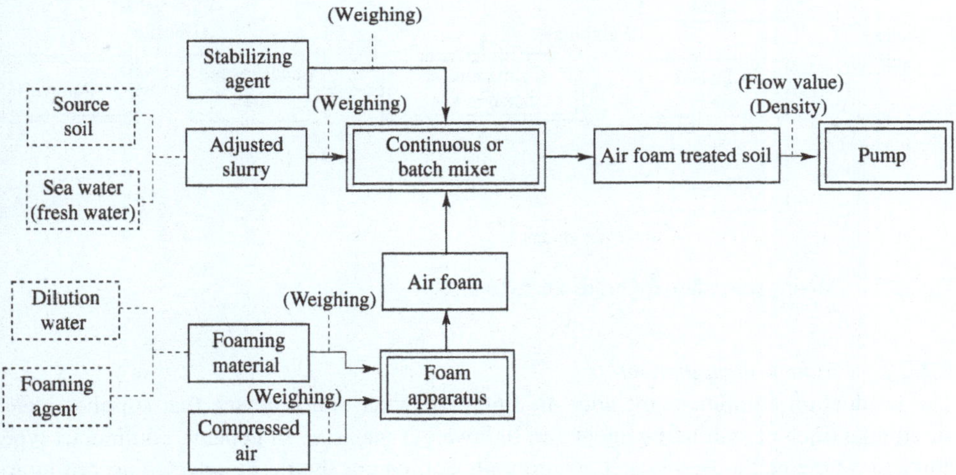

Figure 5.5. Mixing procedure for air foam treated soil.

Figure 5.4 shows an example of a slurrying apparatus. Photograph 5.3 shows a field slurrying process. Photograph 5.3 corresponds to the step in Figure 5.4 where the source soil is placed in a hopper.

5.2.2 Mixing

5.2.2.1 *Mixing process*

Lightweight treated soil is made by mixing a stabilizing agent with the adjusted slurry then adding either air foam or EPS beads to obtain the specified qualities. Because the quantity of lightweight material varies according to the air foam treated soil and the beads treated soil process, these are explained separately below.

- *Air foam treated soil.* The mixing procedure for air foam treated soil is shown in Figure 5.5. Photograph 5.4 shows the air foam that is produced.
- *Beads treated soil.* Figure 5.6 shows the mixing procedure for beads treated soil.

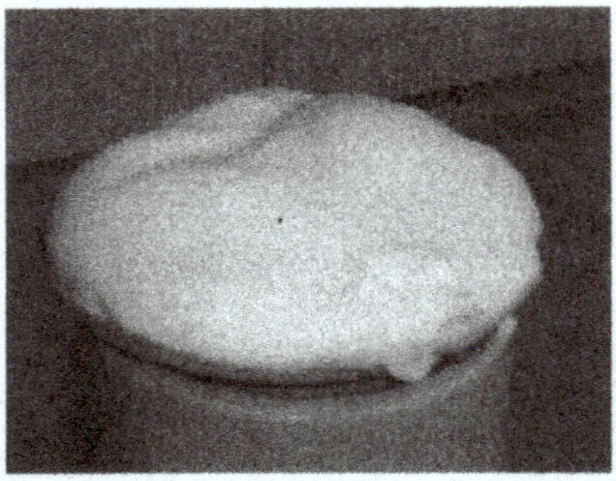

Photograph 5.4. Air foam.

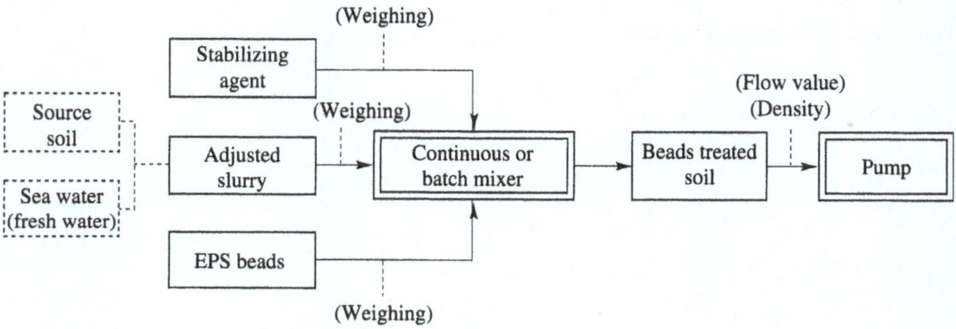

Figure 5.6. Mixing procedure for beads treated soil.

5.2.2.2 *Production equipment*

The production equipment includes an agitating mixer and a device that supplies measured quantities of stabilizing agent and lightweight material. In general, continuous type and batch type methods are used. Appropriate equipment should be selected according to manufacturing capacity. In case of large-scale production in a coastal area, a specialized offshore plant is used and the equipment is carried on a barge.

- *Quantity measuring supplier device for stabilizing agent and lightweight material.* This kind of device measures and supplies the required quantities of the stabilizing agent from a silo. For measuring the quantity of stabilizing agent, in general, volume measuring is used for the continuous type, and weight measuring is used for the batch type. Figure 5.7 shows an example of a device that supplies a measured quantity of stabilizing agent.

  Because EPS beads are extremely light, the quantity is measured by the volume method (continuously supplied by a rotating screw feeder). In case of small-scale production, they are supplied in bags of fixed volume.
- *Mixer.* There are batch type and continuous type mixers, and the type is selected according to the required mixing capacity and performance. The mixing time is determined based on trial mixing.

In case of air foam treated soil

After mixing the adjusted slurry with stabilizing agent in a mixer, the air foam is supplied to the mixer to be kneaded and mixed. Another method is to mix the adjusted slurry and

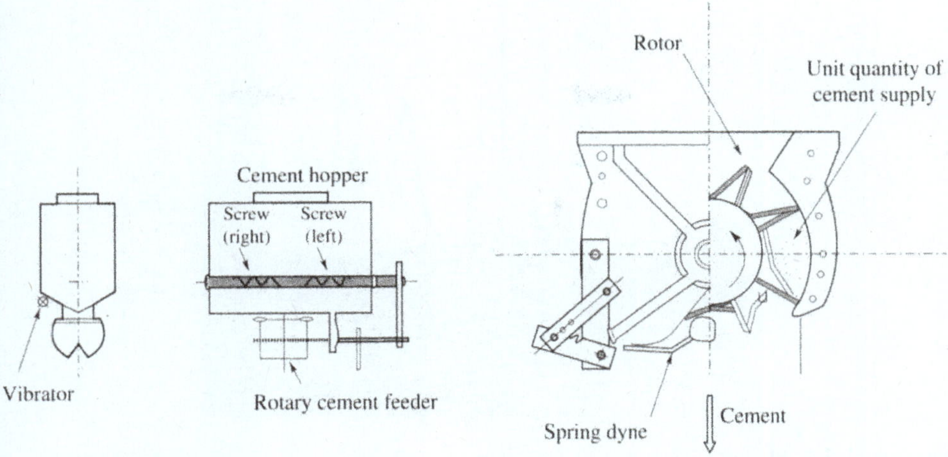

Figure 5.7. Example of a quantity measuring supplier device for stabilizing agent.

stabilizing agent in the mixer, and then to add the air foam in an air pressure pipe for feeding mixed by a blender. Horizontal shaft mixers are often used for the mixing process, due to the large difference between the specific gravities of the materials. However, this must be done carefully, because if the retention time in the mixer is too long, the density becomes inconsistent due to air entrainment.

In case of beads treated soil

Beads treated soil is produced by adding the EPS beads and stabilizing agent to the adjusted slurry and mixing and kneading these materials in the mixer.

5.2.3 Placing

Lightweight treated soil can be placed in the atmosphere and underwater. When lightweight treated soil is placed underwater, its placing rate and method (tremie pipe, pump pouring) should be selected considering its flow and separation properties, and the placing should be carried out giving attention to the placing sequence. The placing process is shown in Photograph 5.5.

5.2.3.1 *Placing process*

- *Procedure using tremie pipe.* The following is the procedure to place the material using a tremie pipe.
  - (i) The range of placing by a single tremie pipe is determined by its covering area (see Figure 5.8).
  - (ii) The tremie pipe is filled with lightweight treated soil.
  - (iii) In each step, placing by a tremie pipe is started at the location where the pipe is always buried in the lightweight treated soil placed at the earlier placing.
  - (iv) After a single cycle of placing the lightweight treated soil, the tremie pipe is raised to a level high enough to avoid disturbing the materials that have already been placed, and is moved horizontally. After it reaches the next point of placing, steps [ii] and [iii] are repeated to place the lightweight treated soil.

Photograph 5.5. Placing.

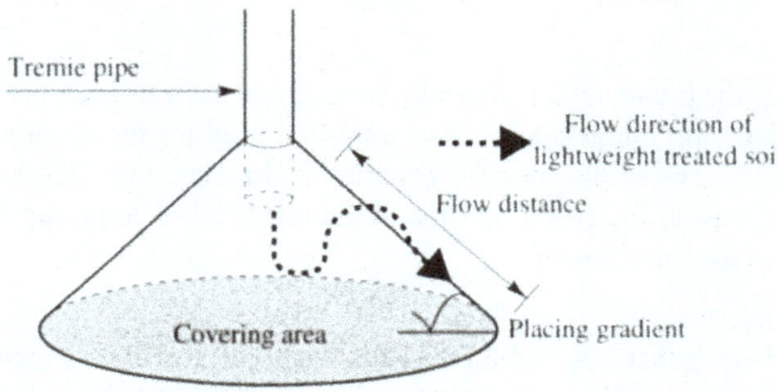

Figure 5.8. Gradient of placed lightweight treated soil by vertical placing.

- *Procedure for direct placing by pumping* (Fig. 5.9a). The following is the procedure used to place the material directly with a pump.
  - (i) The valve on the tip of the pipe filled with lightweight treated soil is closed, then the tip of the pipe is guided to the specified depth.
  - (ii) After starting to operate the pump, the valve on the tip is opened, and the lightweight treated soil is fed at a constant rate.
  - (iii) The lightweight treated soil is placed with the tip of the outlet at a fixed elevation, therefore it remains constantly buried in the lightweight treated soil.
  - (iv) After a single cycle of placing lightweight treated soil, the pump is stopped, then the valve on the tip is closed. The tip is moved to the next location, and the lightweight treated soil is placed by repeating steps [i] to [iii].

Another method is horizontal placing carried out by placing the lightweight treated soil while moving the outlet in the direction opposite to the discharge at an appropriate speed (Fig. 5.9b).

Vertical placing method

Because lightweight treated soil spreads over 360° during placing and pushes the material already placed, thus forming cracks, it becomes immersible.

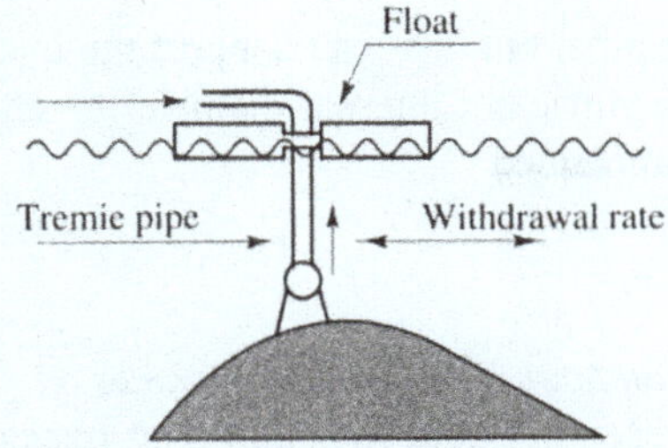

Figure 5.9a. Pump placing methods.

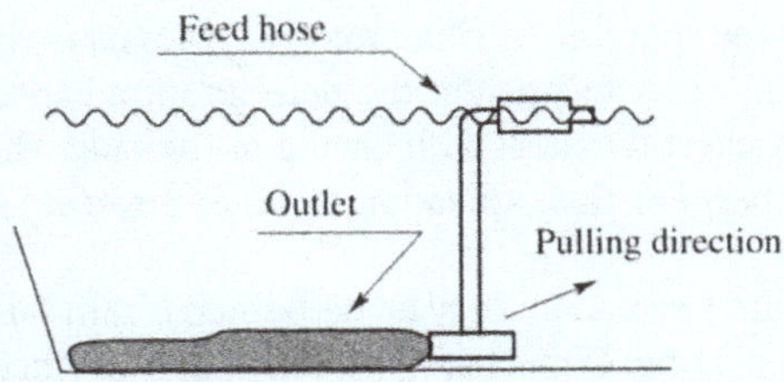

Figure 5.9b. Pump placing methods.

Horizontal placing method
Because lightweight treated soil is placed while the outlet is moved horizontally in the direction opposite to the discharge, it has little effect on the material that has already been placed, but joint traces are formed.

5.2.3.2 *Placing system*
Compression pump
The pump specifications are established considering the lift of the compression pump, the application quantity, etc. and a relay pumping yard is installed when the material is to be pumped for a long distance.

- *In case of air foam treated soil.* Because the pumping distance of air foam treated soil is affected not only by the pump feed capacity, but also by defoaming and the flow value, a squeeze type pump with low pressure pulsation is used to feed air foam treated soil.
- *In case of beads treated soil.* Beads treated soil is fed by squeeze type pumps, piston pumps, etc.

Pipes
The specifications of the pipe are established considering its pressure resistance and the quantity of material to be fed. The pipe arrangement should be decided regarding the range of the application and the locations of the slurrying and the mixing system.

Placing equipment
Lightweight treated soil transported by pumping is placed by a tremie pipe taking into account the separation of materials. When beads treated soil is placed, additional equipment may be necessary to recover the EPS beads separated during the placing.

When selecting a tremie pipe, one that can support the quantity of lightweight treated soil for placing, and with a sufficient internal diameter to smoothly feed the treated soil without separation, is recommended.

5.2.4 Curing

5.2.4.1 *Treatment of section boundaries and joint traces*

It is necessary to treat the section boundaries and joint traces for the lightweight treated soil which is to be placed next to that placed in the previous stage.

5.2.4.2 *Surface treatment*

Curing accompanied by the spraying of water may be necessary when the placing is carried out in the atmosphere, in order to prevent the deterioration caused by rapid drying. It is essential to be extremely careful to deal with drying or freezing when the material is placed during cold or hot weather. For this, spreading sand or covering the material with sheets may be used.

The air foam in air foam treated soil may be defoamed if rain falls on the material during or immediately after the placing. If this happens, the poured surface should be covered with a sheet and the area should be drained to prevent it from becoming submerged.

5.3 CONTROL METHODS

5.3.1 Application control

It is necessary to control the application of lightweight treated soil by the processes shown in Table 5.1.

5.3.1.1 *Material control*

Source soil

The following tests are carried out to investigate the physical properties of the source soil. If necessary, the tests are carried out during execution.

- Soil particle density test (JIS A 1202)
- Soil water content test (JIS A 1203)
- Soil liquid limit and plastic limit test (JIS A 1205)
- Soil particle size distribution test (JIS A 1204)
- Soil wet density test (JGS T 191)
- Soil pH test (JGS T 221)
- Soil ignition loss test (JGS T 221)
- Soil organic carbon content test (JGS T 231)

If the results of these tests indicate that the properties of the source soil differ greatly from those of the soil used for the preliminary mix proportion test, the mix proportion test is performed again to promptly correct the mix proportion of the lightweight treated soil.

If the source soil contains a large quantity of organic material or if the mix proportion test of air foam treated soil might be affected by defoaming, the quantity of air foam is initially set higher than necessary.

Table 5.1. Application procedure–control relationships.

	Dredging and transport	Slurrying	Mixing	Placing	Curing	Completion
1. Material control	Source soil		Lightweight material/stabilizing agent			
2. Slurrying control		Control of the quantity of water added and the density of the adjusted slurry				
3. Mixing control			Measuring the materials placed in the mixer			
4. Placing control				Safety control and environmental control		
5. Completed form control				Depth sounding		Measurements
6. Quality control			Density, flow value, etc.	Density, flow value, etc.		Post-survey (boring)

Stabilizing agent

The type and quantity of stabilizing agent are determined based on the results of the preliminary mix proportion test. When stabilizing agent is delivered, its type, delivery date, quantity, etc. are recorded, then it is stored in a silo. If the stabilizing agent is delivered in bags, it is stored in the order it is delivered in a place where weathering of the stabilizing agent can be prevented.

Lightweight material

- *Foaming agent.* The foaming agent should not be stored where it is exposed to direct sunlight, or stored where it could freeze or be degraded. Also, it is not stored for a long time after it has been opened.
- *EPS beads.* Because EPS beads are a super-lightweight artificial material, they must be stored in accordance with the following precautions:

  In a place with no flames;
  Either in bags or in a silo to prevent them from being scattered;
  In a place where it will not become wet due to rain.

Water

Fresh water should used as the mixing water.

5.3.1.2 *Slurrying control*

It is most important to maintain the quality of the lightweight treated soil during the adjustment of water content and density at the slurrying stage. Therefore, the objects of slurrying control are the water content and density of the material. An γ-ray densitometer, etc. are used to measure the density. To maintain a constant water content, adding water is an effective method if excessive fluctuation of the water content may occur due to the effects of evaporation of water caused by direct sunlight or dilution by rainfall. The adjusted slurry is agitated by an underwater mixer, agitator, etc. in order to prevent the settling of soil particles.

5.3.1.3 *Mixing control*

During mixing, the quantity of materials put in the mixer is controlled and the density and the flow value of the lightweight treated soil are also controlled at the specified levels.

In case of air foam treated soil

- *Adjusted slurry.* The flow rate of the adjusted slurry supplied from the slurry storage tank is measured as it is supplied to the mixer.
- *Stabilizing agent.* The stabilizing agent is weighed and supplied to the mixer from the silo.
- *Air foam.* The foaming agent is diluted with dilution water at a dilution multiplier set by the foaming agent maker to make foaming material. Then the air foam is made by using a foaming machine to foam the foaming material with compressed air at a foaming multiplier set by the foaming agent maker. The quantities of foaming material and compressed air are controlled during the air foam production process.

In case of beads treated soil

- *Adjusted slurry.* The flow rate of the adjusted slurry from the slurry storage tank is measured as it is supplied to the mixer.
- *Stabilizing agent.* As in the air foam case, the stabilizing agent is weighed and supplied to the mixer from the silo.
- *EPS beads.* Because EPS beads are extremely light, the volume method is used for measuring (continuous supply by the rotation of a screw feeder).

5.3.1.4 *Placing control*

Control of the quantity and the section of placing

Lightweight treated soil is placed immediately after it is made and the placing quantity is controlled by a flow rate gauge. Photograph 5.6 shows an electromagnetic flow rate gauge and Photograph 5.7 shows an indicator panel of an electromagnetic flow rate gauge. The placing section is controlled with records such as drawings.

Control of effects on surrounding structures

Immediately after it has been placed, lightweight treated soil behaves like a fluid, but its strength increases with elapsed time. There are cases where surrounding structures and ground are destabilized by its behavior as a fluid body. In such cases, placing control is performed considering the effect of the placing of the lightweight treated soil on the surrounding structures and the ground. Water level is controlled appropriately during underwater placing.

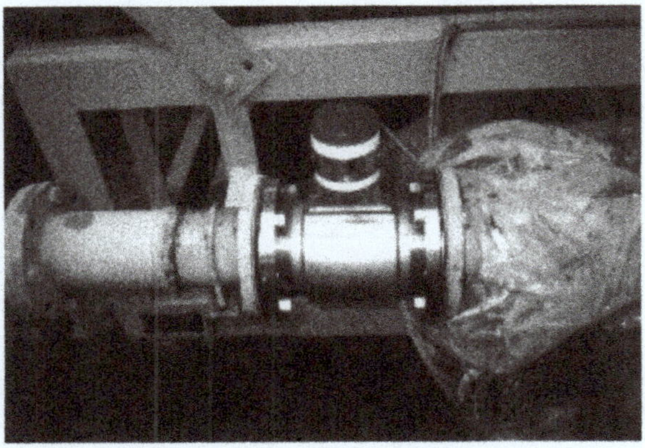

Photograph 5.6. Electromagnetic flow rate gauge.

Photograph 5.7. Electromagnetic flow rate gauge indicator panel.

Control of the marine environment

Lightweight treated soil is applied considering its effects on the ocean and other aspects of the environment. Neutralization treatment is carried out to deal with alkaline water produced by the cleaning fluid used to clean the equipment. Because EPS beads used to make beads treated soil may separate and float to the surface, these EPS beads are recollected and reused.

During the construction, *pH*, COD, turbidity, *SS* and other water qualities are checked as necessary in order to prevent environmental problems.

5.3.2 Completed work control

To control the completed work, the area where the material was placed is measured with an echo sounder or sounding lead, or manually by a diver. In the case of completed work control immediately after placing while lightweight treated soil is in the fluid state, it may be impossible to accurately measure the boundary with the water. In such cases, the shape of the completed work is measured on the day after pouring.

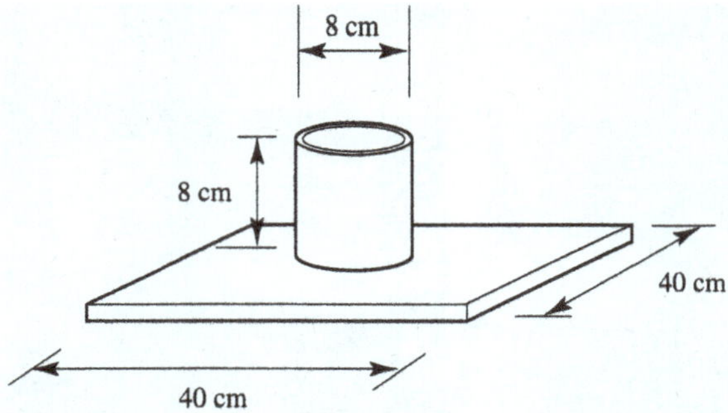

Figure 5.10. Flow test.

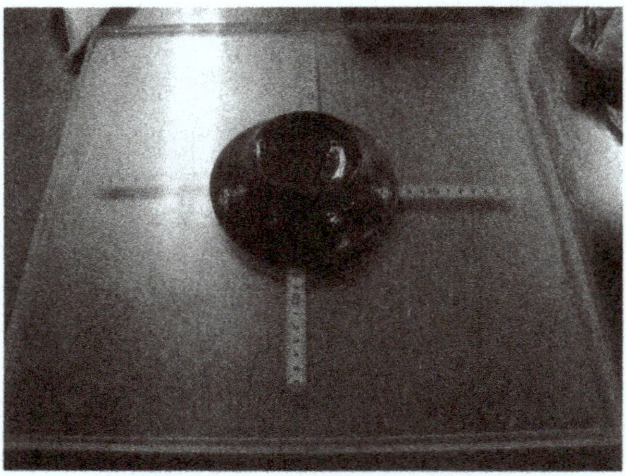

Photograph 5.8. View of a flow test.

5.3.3 Quality control

5.3.3.1 *Quality control of mixing work and placing work*

The following tests must be performed during mixing and placing to confirm that the quality of the lightweight treated soil satisfies the established control values determined by the results of laboratory mix proportion tests.

Flow value

The flow value is measured in accordance with the cylinder method using a cylinder with an internal diameter of 8 cm and height of 8 cm based on the flow tests (JHS A 313) (see Fig. 5.10). Photograph 5.8 shows an example of the test.

Density

The density of a lightweight treated soil that has been mixed is measured by hand balances or vessels with fixed capacity, or by automatic densitometers (see Photograph 5.9).

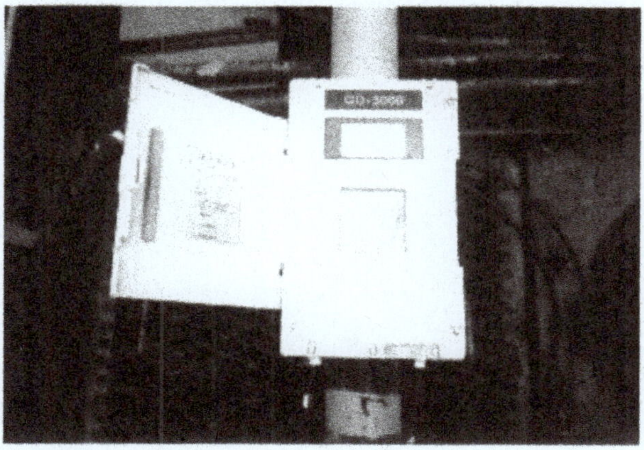

Photograph 5.9. γ-ray densitometer.

Air content

The air content of lightweight treated soil that has been mixed is examined by methods such as alcohol measurement.*

Measuring the density, strength, and water content using a prepared specimen

Density and unconfined compressive strength tests and measurements of the water content of the lightweight treated soil after stabilization are carried out using specimens aged for 28 days to confirm that the stipulated values are all satisfied. Specimens of 7 days aging are also tested, if necessary.

5.3.3.2 *Post work surveys*

Post work surveys are carried out to confirm the long-term quality of lightweight treated soil at construction sites and to study the underwater strength/atmospheric strength ratio. In this case, the material is sampled by a check boring and the following tests are performed.

Density

The densities of the samples are measured simultaneously with unconfined compressive strength tests.

Unconfined compressive strength

Unconfined compressive strength tests for specimens of 28 days of aging are performed to confirm that the unconfined compressive strength satisfies the specified values. The strength of younger materials and of older materials are also examined, if necessary.

Water content

The water content of the samples is measured at the same time as their unconfined compressive strength is tested.

---

* 200 m*l* of a sample is placed in a 500 m*l* measuring cylinder, 200 m*l* of water is added, the mixture is thoroughly shaken to separate the foam, then 100 m*l* of alcohol is dripped into it to completely defoam it. Next, the scale on the measuring cylinder is recorded and the air content is calculated by the following equation.
Air content (%) = (specimen (200 m*l*) + water (200 m*l*) + alcohol (100 m*l*) − reading after defoaming (m*l*))/specimen (200 m*l*).

Chapter 6

Examples of Field Application

6.1 CASE STUDIES IN JAPAN

The first use of lightweight treated soil as a ground material in the history of port or airport construction projects in Japan was for quaywall improvement work in the Port of Fushiki-Toyama in 1992 [1],[2]. The quaywall was a wharf type structure, but during the nearly 30 years since its construction, the corrosion of its steel pipe piles and steel sheet piles had caused it to deteriorate to such a degree that it could no longer satisfy the allowable stress during an earthquake. Therefore, as shown in Figure 6.1, the quaywall was reconstructed and air foam treated soil was used as a backfill material in one part of the reconstructed section. This material was used in order to reduce the earth pressure behind the quaywall and to secure the stability at the front of the quaywall under the action of waves. It was a small-scale work with only approximately 900 m^3 placed. Photograph 6.1 shows a view of the placing. Approximately one year after the construction, some specimens were sampled to study its density and strength. The results showed that the strength and density satisfied the design conditions, as shown in Figure 6.2, thus verifying the applicability of the method.

Lightweight treated soil was fully used for a large-scale project for the first time as a part of the earthquake disaster recovery project in Port Island of Kobe Port in 1996. The surface ground behind a quaywall that was being reconstructed to repair the damage caused by the Hanshin-Awaji Earthquake, was replaced with air foam treated soil to reduce the earth pressure behind the quaywall and to secure its earthquake stability. The source soil was dredged clays from the shipping route in Kobe Port. It was the first large-scale

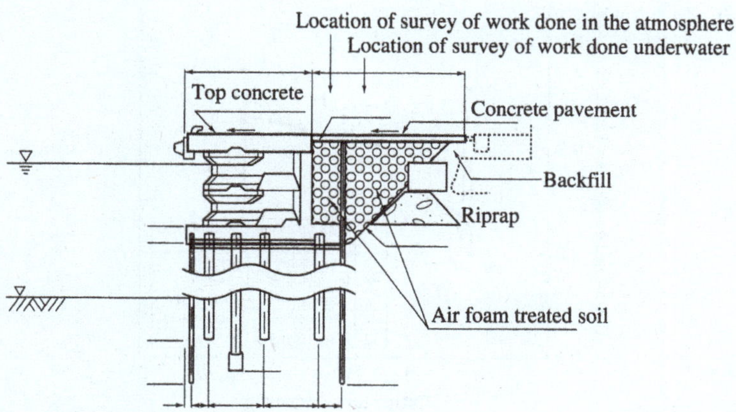

Figure 6.1. Standard section.

Photograph 6.1. Placing at Fushiki-Toyama.

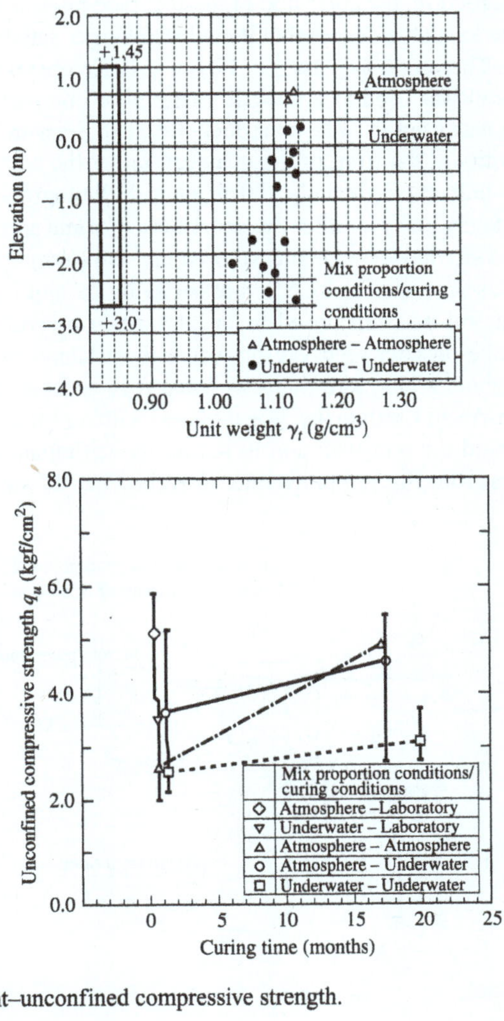

Figure 6.2. Unit weight–unconfined compressive strength.

Table 6.1. List of previous cases of application.

No.	Work name	Execution period	Source soil	Treatment method	Quantity placed (m^3)
1	Fushiki-Toyama Port quaywall improvement	1992.11–1993.1	Soil excavated at the site	Air foam	900
2	Port Island quaywall reconstruction	1995.12–1996.5	Soil dredged from a channel in Kobe Port	Air foam	21,610
3	Tokyo International Airport Case 1	1996.5–1996.6	Surplus soil from shield work	Air foam EPS beads	1,930
4	Kumamoto Port, field test of SGM lightweight treated soil	1998.5–1998.8	Soil dredged from a channel in Kumamoto Port	Air foam EPS beads	890
5	Tokyo International Airport Case 2	1998.9–1998.12	Construction waste soil and soil excavated at the site	Air foam	32,220
6	Quaywall improvement work at the new harbor at Ishigari	1998.9–1999.3	Soil excavated at the site	Air foam	3,240
7	Tokyo International Airport Case 3	1999.2–1999.3	Construction waste soil	Air foam	11,020
8	Tokyo International Airport Case 4	1999.2–1999.3	Construction waste soil	Air foam	8,400
9	Quaywall improvement work at the new harbor at Ishigari	1999.6–2000.1	Soil excavated behind a quaywall	Air foam	4,114
10	Reinforcement of a revetment at Oi Wharf in Tokyo Port	1999.8–2000.4	Soil excavated at a wharf	Air foam	10,210
11	Tokyo International Airport Case 5	1999.9–1999.11	Excavated soil Construction waste soil	Air foam	23,490
12	Improvement of a quaywall in Yokohama Port	2000.3–2001.3	Dredged soil	Air foam	43,170

use of the method, with the work area extending for 190 m and involving the placing of about 20,000 m^3 of lightweight treated soil. The following year, lightweight treated soil was used to prevent settlement and reduce the earth pressure behind a revetment constructed at Tokyo International Airport. At Tokyo International Airport, lightweight treated soil made with construction waste soils and leftover soils was used for parallel taxiways, peripheral roads, and runways, with the total amount of application at the site reaching approximately 80,000 m^3. In 1998, a field test of underwater placing was carried out in deep water of approximately 10 m by a full-size plant in Kumamoto Port [6]. It was confirmed that even in deep water, it is possible to obtain adequate quality: unit weight of 12 kN/m^3 and unconfined compressive strength of 200 kN/m^2. Table 6.1 lists previous application cases and Figure 6.3 indicates locations where lightweight treated soil has been used. As of April 2000, a total of approximately 120,000 m^3 of lightweight treated soil has been used for 11 projects. Its

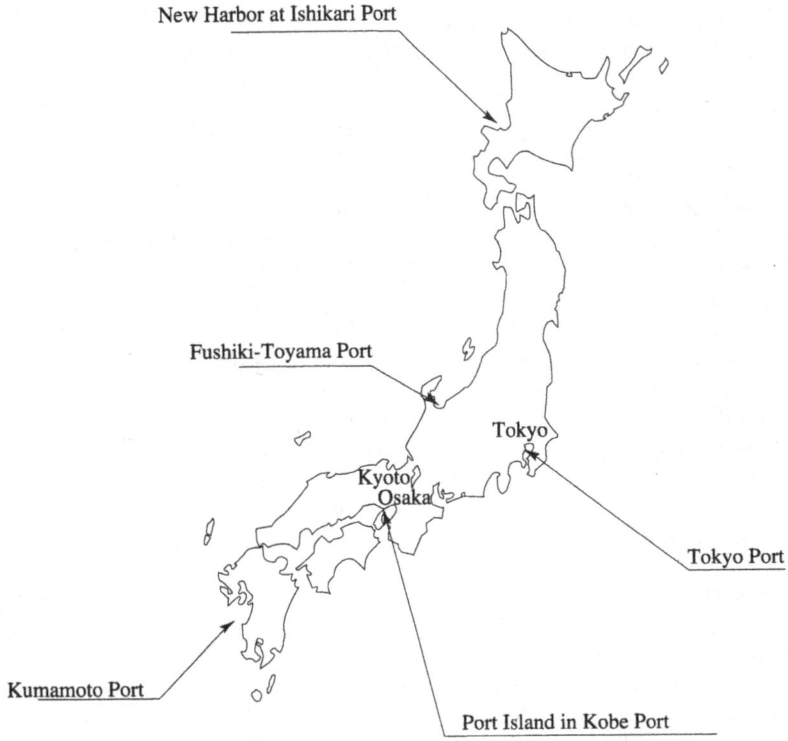

Figure 6.3. Locations where lightweight treated soil has been applied.

scope of application is no longer limited to its initial use as backfilling for quaywalls; it has been used to prevent settlement of shield tunnels and to prevent liquefaction, and it is expected to be used for even more purposes in the future. In most of these application cases, air foam treated soil was used. Only about 2,000 m^3 of beads treated soil have been used at two locations – at the revetment work site at Tokyo International Airport and in the field test in Kumamoto Port – and both cases were trials. This chapter presents detailed descriptions of two representative applications of lightweight treated soil: the work on Port Island in Kobe Port and the work at Tokyo International Airport.

6.2 EXAMPLE: PORT ISLAND IN KOBE PORT [3][4][5]

The restoration of harbor facilities in Kobe Port that were damaged by the Hanshin-Awaji Earthquake Disaster of 1995 was carried out using various methods suitable for each facility considering the degree of damages and the restricted conditions of each facility. At quaywalls where the deformation was relatively small and the post-damage normal line retained a certain degree of linearity, the restoration work was generally carried out by reducing the earth pressure that would act on the back of the quaywall during an earthquake. Methods used to reduce earth pressure during earthquakes that were adopted for the restoration project included improvement of the ground behind the quaywall to increase its strength,

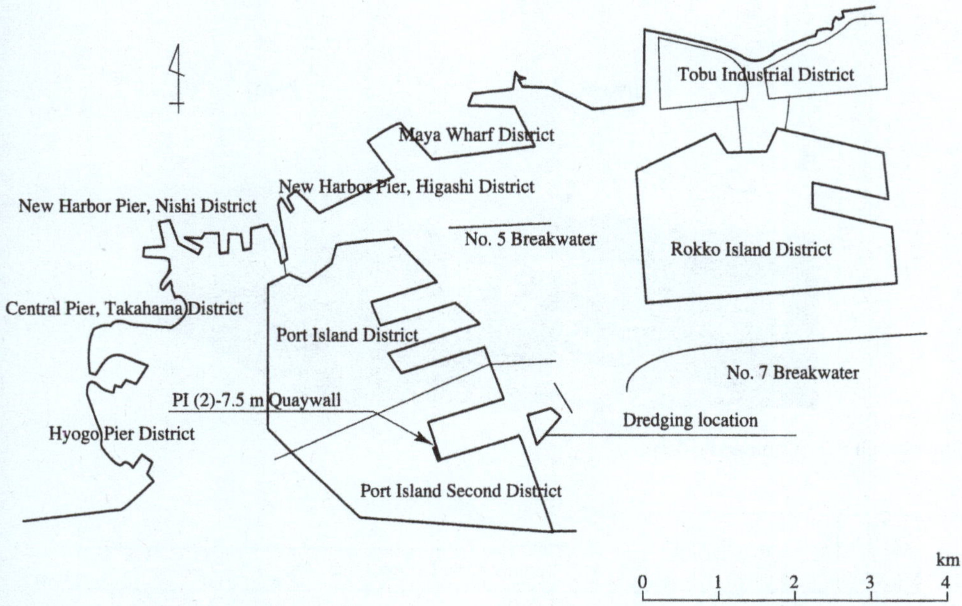

Figure 6.4. Work site location.

weight saving of backfills behind the quaywall, and ground excavation behind the quaywall to structurally reduce its earth pressure. The method used to secure the earthquake stability of the quaywall in Section No. 2 on Port Island in Kobe Port by reducing the earth pressure, was the replacement of the ground surface behind the quaywall with air foam treated soil. Figure 6.4 shows the facility where this work was done – the Section No. 2 quaywall on Port Island – and its location.

The quaywall is 183 m in length, and the water depth gradually becomes deeper from 7.5 m to 15.0 m as it approaches the 15-m deep quaywalls to the south. At the time of the earthquake, this facility was under construction, at the stage of completing the concrete capping of filled caissons and backfilling.

The earthquake pushed out the normal line of the quaywall by between 0.8 and 3.8 m, lowered the tops of the caissons by between 1.1 and 1.2 m, and inclined the caissons by between −2.5° and +3.0°. The horizontal displacement was relatively large towards the north and the caisson tops were submerged. Photograph 6.2 shows the damages to the caissons and Figure 6.5 shows the damage conditions. The caissons deformed by the damage were used directly without replacement in order to shorten the time required for reconstruction and to match the normal line with those of adjoining quaywalls. However, because the caissons experienced settlement of 1.1 and 2.5 m, backfilling with ripraps or ordinary soils was not appropriate due to the high earth pressure that they generate. Therefore, foundation ripraps were added in front of the caissons to support their passive resistance, and the ground behind the quaywall was replaced with air foam treated soil to reduce the earth pressure on the back of the walls. Figure 6.6 is a section diagram of the restoration.

The design density of the air foam treated soil was 1.0 t/m^3 for the placing in the atmosphere (H.W.L. +1.7 m and shallower) and was 0.2 t/m^3 in the underwater placing section (elevation +1.7 m and deeper). According to the calculation based on the slice method,

Photograph 6.2. Damaged caissons.

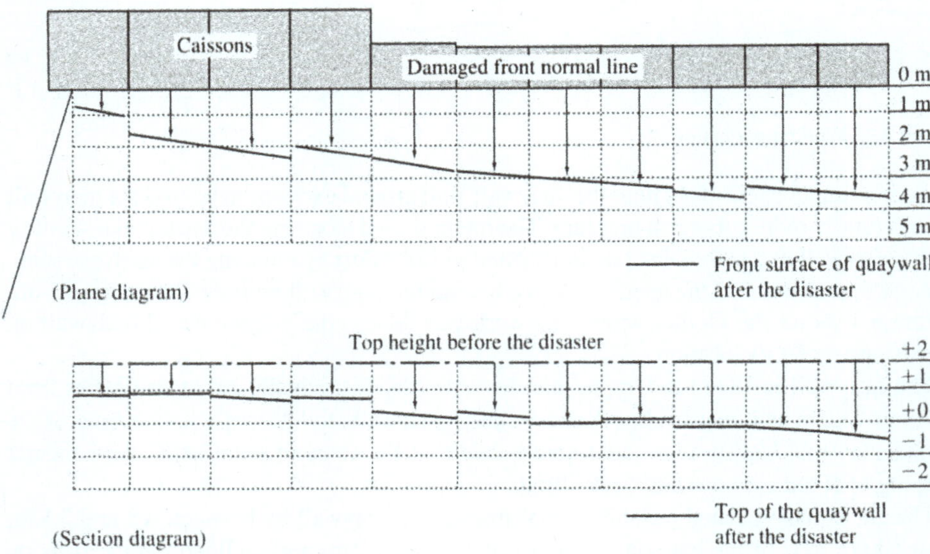

Figure 6.5. Caisson damage.

it was confirmed that the active earth pressure during an earthquake was lowered by about 30% to satisfy the design values by lightening the ground of 4.5 m in thickness from an elevation +3.5 m to −1.0 m and of 30 to 40 m in width on the ground surface. The design strength of the air foam treated soil was established at the unconfined compressive strength of 200 kN/m^2 considering the fact that this ground is an apron road surface and part of it is the support ground for the foundation of cranes.

The source soil was dredged from the shipping routes of Kobe Port for the restoration work. Its properties are shown in Table 6.2.

The stabilizing agent was blast-furnace slag cement type B, the foaming agent used was a surface-active agent with 10× dilution and 25× foaming rates. From the results of

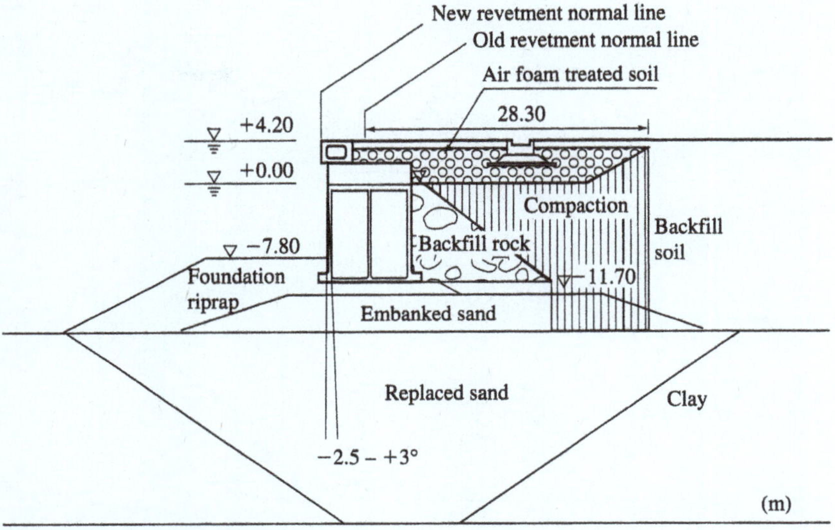

Figure 6.6. Section diagram of the restoration.

Table 6.2. Properties of the source soil.

Soil particle density (g/cm^3)	2.709
Soil content (%)	
Gravel	3
Sand	14
Silt	54
Clay	28
Water content (%)	122
Liquid limit (%)	97
Plastic limit (%)	41
Plasticity index I_p	56
Wet density (g/cm^3)	1.39
Ignition loss (%)	8.8
Organic material content (%)	3.02
pH	8.1

Table 6.3. Mix proportion specifications.

	Adjusted slurry (kg/m^3)	Density at placing (g/cm^3)	Stabilizing agent (kg/m^3)	Air foam (l/m^3)	Flow value (mm)
Atmosphere	849	1	140	279	150–200
Underwater	952	1.1	140	196	150–200

mix proportion tests with the target laboratory mix proportion strength (aging of 28 days) determined at 600 kN/m^2, which is three times the design strength, the mix proportion was decided at 140 kg of cement, 279 liters of air foam for the atmospheric section, and 196 liters of air foam for the underwater section per m^3 of air foam treated soil. The mix proportion specifications are shown in Table 6.3. However, because the in-situ placing test revealed

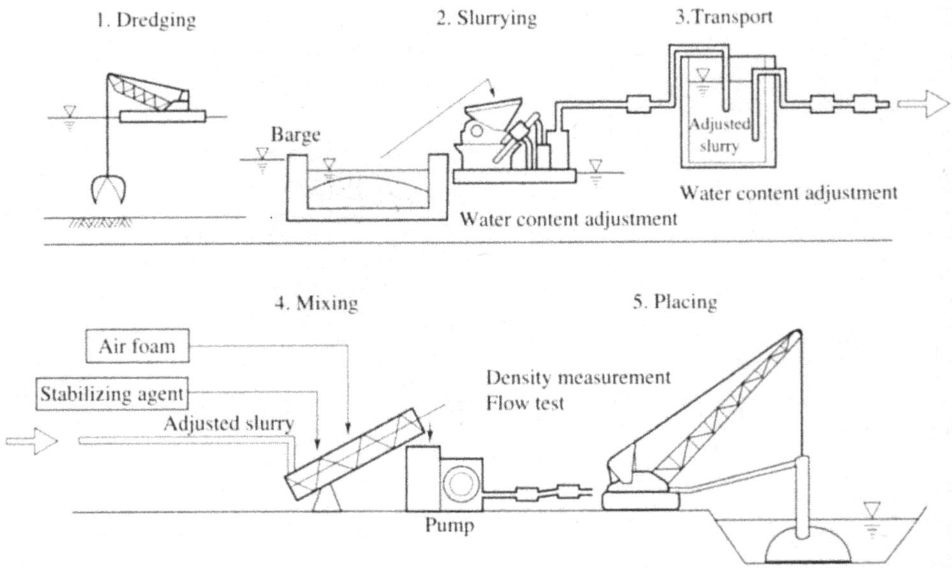

Figure 6.7. Application system.

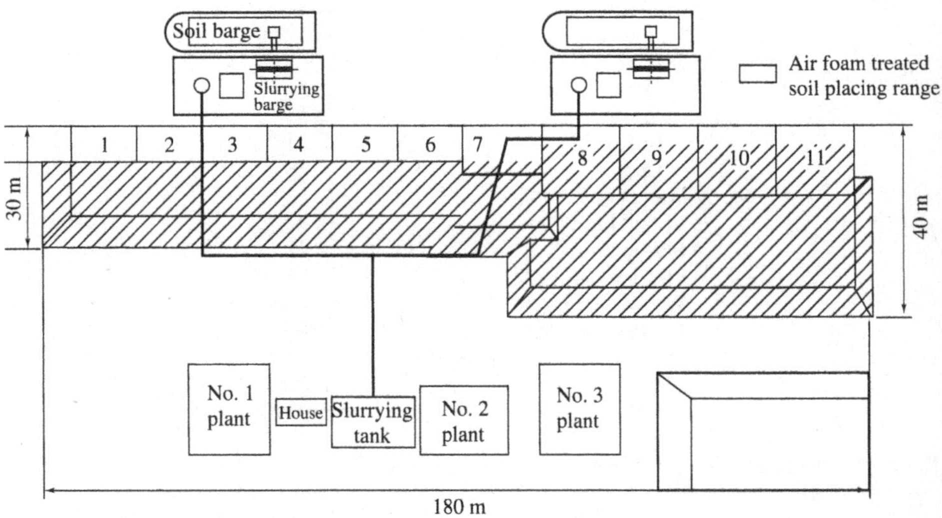

Figure 6.8. Plant layout diagram.

that some of the foam would be defoamed during mixing, pressure feeding, and placing, the actual quantity of air foam was increased by between 15% and 20% larger than the value determined in the field application. Figure 6.7 shows the application system used for this work and Figure 6.8 shows the plant layout. Table 6.4 lists the principal machines used.

Quality control of the application included the flow value, wet density, water content, and unconfined compressive strength of the air foam treated soil. For the efficiency of work and the quality control, γ-ray densitometers were installed at each stage of the application

Table 6.4. List of principal machines.

Machine name	Model/capacity	Number	Remarks
Barge	1500 t	1	Slurrying barge
Back hoe	1.2 m^3	1	Slurrying barge
Vibrating screen	2.2 kW	2	Slurrying barge
Agitator	22–15 kW	4	Slurrying barge
Slurry pump	110 kW	1	Slurrying barge
Underwater pump	5.5, 2.2 kW	2	Slurrying barge
Generator	350 kVA	1	Slurrying barge
Relay slurry tank	30 m^3	1	Relay slurry tank
Slurry feed pump	22 kW	3	Relay slurry tank
Generator	200 kVA	1	Relay slurry tank
Mixing machine	40 m^3/h	6	for 3 pumps
Foamer	700 *l*/min	6	for 3 pumps
Generator	300 kVA	6	for 3 pumps
Cement silo	30 t	6	for 3 pumps
Slurry tank	5–30 m^3	3	for 3 pumps
Agitator	3.7 kW	3	for 3 pumps
Sand pump	3–4B	8	for 3 pumps
Squeeze pump	40–80 m^3/h	6	for 3 pumps
Electromagnetic flow rate gauge		8	All
γ-ray densitometer		6	All

Photograph 6.3. Underwater placing.

system to maintain the specified density. The underwater placing of the air foam treated soil was carried out in water areas behind the quaywall blocked away from the outer sea area. Environmental monitoring was also carried out by taking water samples around the site for water quality tests to confirm that the placing of the air foam treated soil had no harmful effect on the surrounding sea water. Photographs 6.3 and 6.4 show the underwater placing and the atmospheric placing respectively.

Photograph 6.4. Placing in the atmosphere.

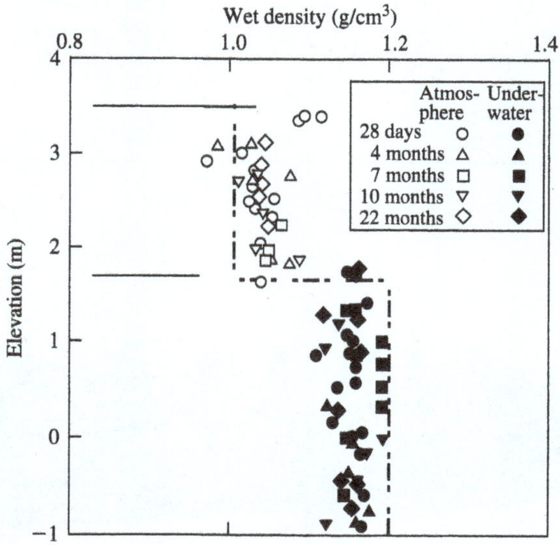

Figure 6.9. Depth distribution of wet density.

To investigate the change of mechanical properties of the air foam treated soil placed under sea water with time, sampling and in-situ tests were carried out at almost the center of the place where the air foam treated soil had been placed at 1, 4, 7, 10, and 22 months after construction.

Figure 6.9 shows the depth distribution of the wet density ρ_t of the specimens. Near the top, the density was a little high, but overall it was generally uniform with little variation. In the section placed in the atmosphere, the value of ρ_t ranged from 0.95 to 1.10 g/cm^3 with a mean value of 1.03 g/cm^3. In the section of underwater placing, the value of ρ_t ranged from 1.10 to 1.19 g/cm^3 with a mean value of 1.15 g/cm^3. These values were a little higher than the design density in the atmospheric section and a little lower in the underwater section,

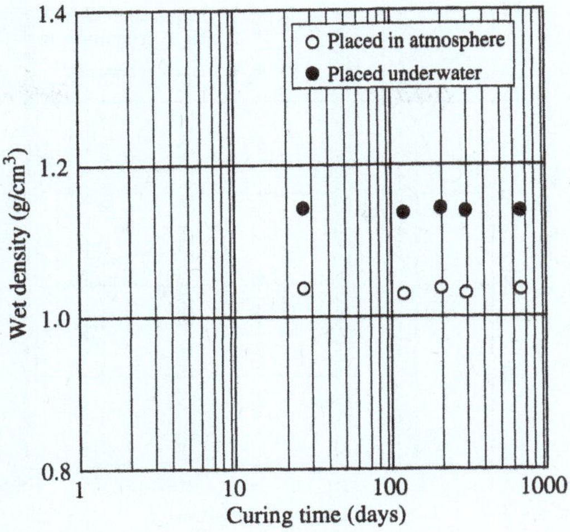

Figure 6.10. Change in wet density over time.

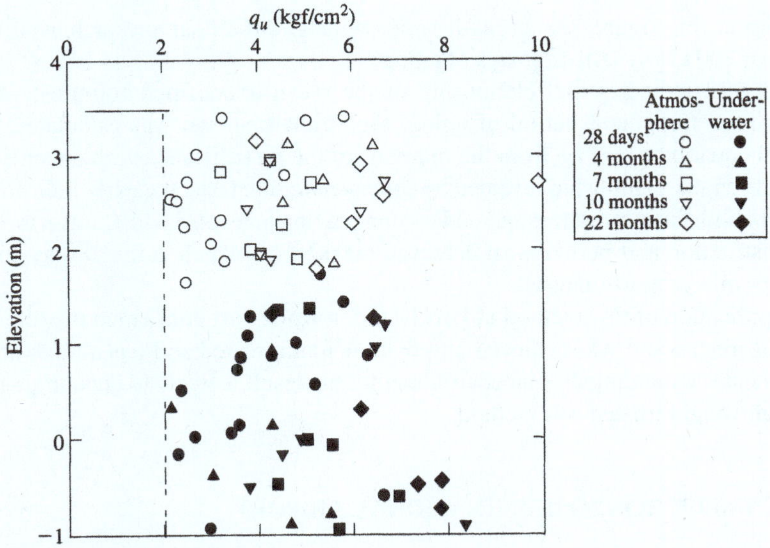

Figure 6.11. Depth distribution of unconfined compressive strength.

but on the whole it shows a rather uniform distribution with little variation. It is assumed that the density had increased a little larger than the design value during the placing in the atmosphere because of the volume shrinkage caused by the initial hardening after placing. Figure 6.10 shows the variation of wet density according to aging. From this figure, it is clear that in both the atmosphere and underwater placing sections, little change had occurred even 22 months after placing.

Next, Figure 6.11 shows the depth distribution of the unconfined compressive strength of the samples taken from the underwater and atmospheric placing sections respectively.

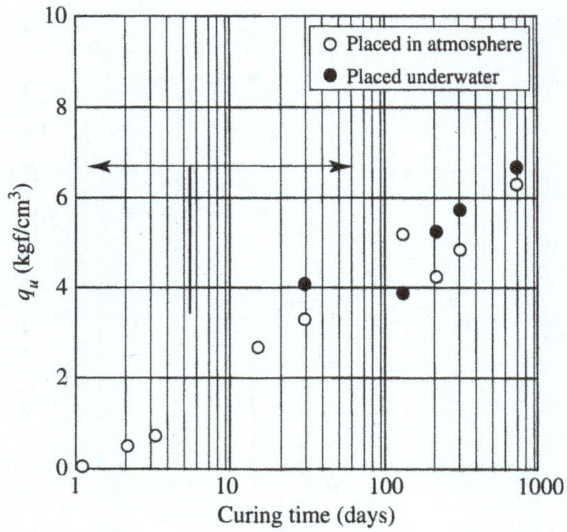

Figure 6.12. Change of unconfined compressive strength.

According to this figure, the strength varies widely, but all samples achieved the target strength of 200 kN/m^2 after 28 days of aging.

Figure 6.12 indicates the relationship of the mean unconfined compressive strength with age. For the young period of aging, the strength shown was calculated using the empirical equation $q_c = 7\,q_u$ from the relations of the tip resistance with unconfined compressive strength relationship revealed by cone penetration tests. It clearly indicates that the strength in both the atmosphere and underwater placing increased with time up to 22 months after construction and both almost achieved 600 kN/m^2, which is the 28-day strength by laboratory mix proportion tests.

The application of this method at Port Island in Kobe Port confirmed that the specified quality of treated soil was achieved and that air foam treated soil has almost no harmful effect on the surrounding marine environment. This result is likely to encourage future use of the lightweight treated soil method.

6.3 EXAMPLE: TOKYO INTERNATIONAL AIRPORT

The Tokyo International Airport Offshore Expansion Project was undertaken to achieve three goals: secure full air transport handling capacity, reduce aircraft noise, and effectively utilize a waste disposal site. The project comprised the utilization of the 468-hectare Tokyo waste disposal site for airport facilities by expanding an additional 341 hectares of waste disposal and land reclamation areas. In the planning and implementation of this project, many problems had to be overcome. First, because reclaimed ground is very soft, differential settlement is anticipated to occur on such ground. Secondly, the work was severely restricted because it had to be performed on the land adjoining the existing airport facilities that were in full operation. Thirdly, reclaimed land preparation work, the construction of a railway tunnel, utility tunnels, and roads, and other works had to be carried out at the same time in the same area, and all of these factors would mutually influence each other. To overcome all

Photograph 6.5. Aerial view of Tokyo International Airport [7].

Table 6.5. Application cases of lightweight treated soil for offshore expansion project.

Case	Purpose of use	Design	Source soil	Soil treatment method	Quantity poured
Case 1	Settlement and earth pressure reduction behind a revetment	$\gamma_t = 11.0\,\text{kN/m}^3$ $q_u = 200\,\text{kN/m}^2$	Shield work surplus soil	Air foam EPS beads	$965\,\text{m}^3$ $965\,\text{m}^3$
Case 2	Tunnel settlement prevention	$\gamma_t = 11.0\,\text{kN/m}^3$ $q_u = 200\,\text{kN/m}^2$	Surplus construction soil and soil excavated at the site	Air foam	$32{,}220\,\text{m}^3$
Case 3	Reduction of earth pressure on an existing box culvert	$\gamma_t = 11.0\,\text{kN/m}^3$ $q_u = 200\,\text{kN/m}^2$	Surplus construction soil	Air foam	$11{,}020\,\text{m}^3$
Case 4	Prevention of tunnel settlement	$\gamma_t = 11.0\,\text{kN/m}^3$ $q_u = 200\,\text{kN/m}^2$	Surplus construction soil	Air foam	$8{,}400\,\text{m}^3$
Case 5	Prevention of tunnel settlement and prevention of liquefaction	$\gamma_t = 11.0\,\text{kN/m}^3$ q_u=200 kN/m^2	Soil excavated at the site, surplus construction soil	Air foam	$23{,}490\,\text{m}^3$

these restrictive factors, advanced technologies were required and some new technologies were introduced in order to complete the entire project.

Lightweight treated soil was one new technology applied to the Tokyo International Airport Offshore Expansion Project. As shown in Table 6.5, a total of 80,000 m^3 of lightweight treated soil was used for five steps in the project between 1996 and 1999. Figure 6.13 shows where it was used. In this project, the scope of application of lightweight treated soil was largely expanded: it was not only used for quaywall backfilling, but also to reduce the overburden pressure on underground structures, to reduce settlement, and to prevent

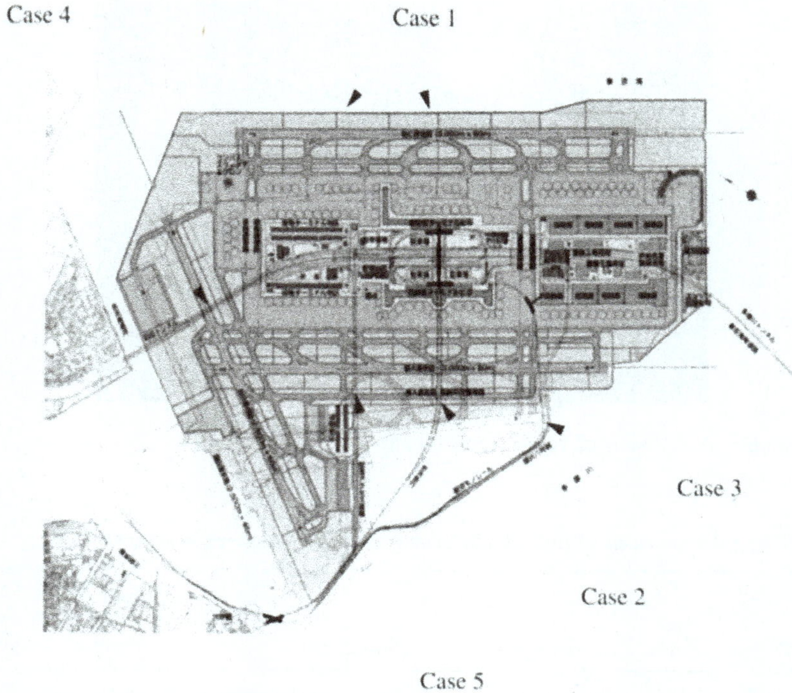

Figure 6.13. Execution locations.

liquefaction. Also, beads treated soil was used for the first time for Case 1 of the construction of the peripheral revetment.

The following section gives a detailed description of Case 1, which was the first use of the method in the Tokyo International Airport Offshore Expansion Project and Case 2, where the method was used to overcome the problem of settlement in the construction of a shield tunnel.

6.3.1 Case 1 [4][8]

The peripheral revetment in the third phase of construction of the Tokyo International Airport Offshore Expansion Project was initially constructed as a revetment to enclose a waste disposal reclamation area. Because it was found that it could not withstand the allowable overflows by high waves, it was decided to raise the level of the embankment. However, it was expected that the use of ordinary pit sand would increase the earth pressure acting on the existing sheet pile quaywall during earthquakes, so a decision was made to use either air foam treated soil or beads treated soil to reduce the earth pressure. Figure 6.14 shows the cross-section of Case 1. Two types of treated soil, air foam and beads treated soil containing EPS beads with a diameter ranging from 1 to 3 mm, were adopted. The total length of the work was 150 m, with 75 m placed by air foam treated soil and the other 75 m placed using the beads treated soil. For both cases, the thickness of treated soil was 2.6 m placed in steps in the atmosphere, and 13 m^3 of lightweight treated soil were placed

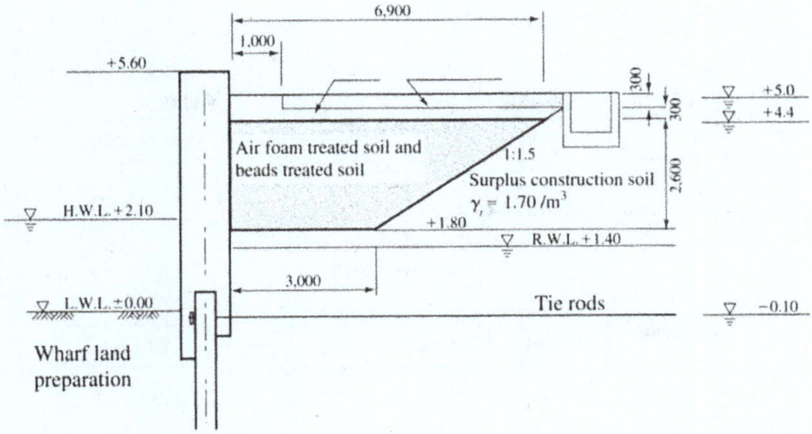

Figure 6.14. Execution section.

Table 6.6. Properties of the shield surplus soil.

Natural water content (%)	Soil particle density (g/cm^3)	Gradation (%)			Liquid limit (%)	Plastic limit (%)	Plastic index
		Sand	Silt	Clay			
39.2	2,693	19	56	25	56.3	22.3	34.0

Table 6.7. Mix proportion specifications.

	Density (t/m^3)	Dry soil (kg/m^3)	Water (kg/m^3)	Stabilizing agent (kg/m^3)	Lightweight material (l/m^3)	Flow value (mm)
Air foam treated soil	1.1	423	571	94	260	Min. 160
Beads treated soil	1.1	478	538	75	268	Min. 120

in each 1 m^2 of cross-sectional area. Approximately 1,000 m^3 of treated soils for each type were used.

The source soil used was construction waste soil produced by the construction of a shield tunnel for a railway running across the airport site. The soil was prepared by adding a coagulant to shield slurry produced by excavation of a hollocene clay layer, then dewatering it with a filter press. Table 6.6 shows the properties of the shield soil. The mix proportion was established to achieve strength of 1.1 g/cm^3 and design unconfined compressive strength of 200 kN/m^2. The stabilizing agent was blast-furnace slag cement type B and the foaming agent used was a surface-active agent with 10× dilution and 25× foaming rates. The EPS beads used had an 80× foaming multiplier (apparent density of 0.0125 g/cm^3). Table 6.7 shows the mix proportion specifications.

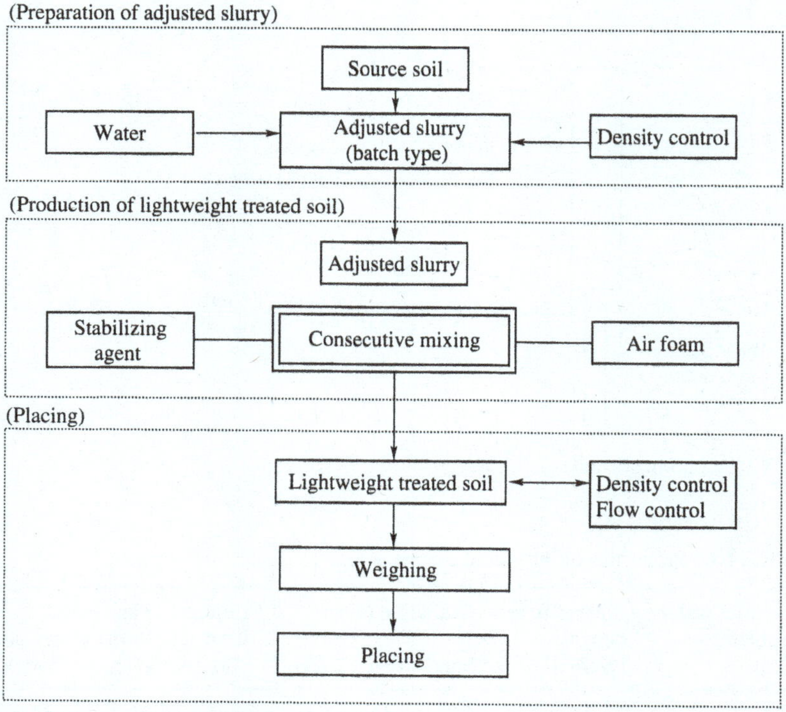

Figure 6.15. Lightweight treated soil application procedure (air foam).

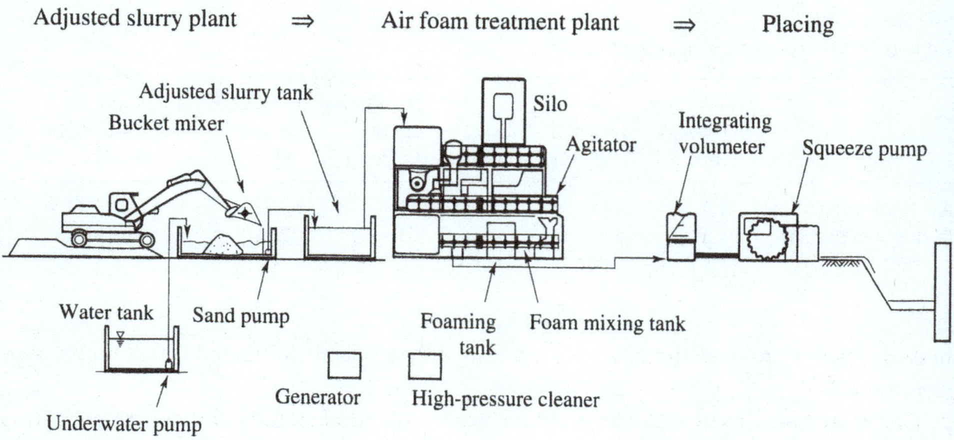

Figure 6.16. Air foam treated soil system.

Different construction systems were adopted separately for the air foam treated soil and the beads treated soil. The construction procedure, systems, and principal machines used for the air foam treated soil method are shown in Figure 6.15, Figure 6.16, and Table 6.8 respectively, while those for the beads treated soil method are shown in Figure 6.17, Figure 6.18, and Table 6.9 respectively.

Table 6.8. List of principal machines.

Machine name	Specifications		Number
Slurrying mixer	15 m^3/H	32.1 kW	1
Back hoe with slurrying device	0.7 m^3		1
Stabilizing agent silo	30 t	12 kW	1
Generator	200 kVA		1
Distributing panel			1
Squeeze pump	PQ · 20	15 kW	1
Integrating flowmeter	IQ With recorder		1
High pressure cleaner		3.7 kW	1
Air compressor	50 PS		1
Sea water tank	30 m^3		1
Slurrying tank	25 m^3		1
Slurry storage tank	20 m^3		1
Fresh water tank	5 m^3		1
Sand pump and underwater pump	4^5–4	3^5–2	1
Engine welder			1
Crane for placing	25 t		1

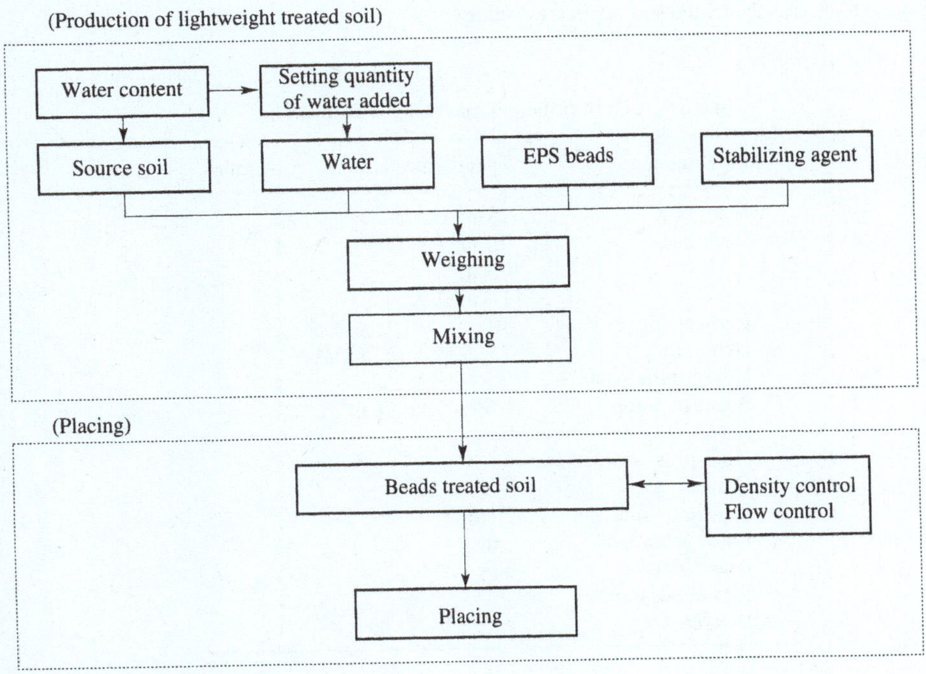

Figure 6.17. Beads treated soil production procedure.

The air foam treated soil application procedure was almost identical to that used in Kobe, but the beads treated soil was placed in 1 m^3 batches instead of continuously. The material was placed to a depth of 2.6 m in three layers (1.0 m, 1.0 m, and 0.6 m layers) as the tip of the ejector moved steadily to place the material flat. It took 9 days to place 1,000 m^3 of the

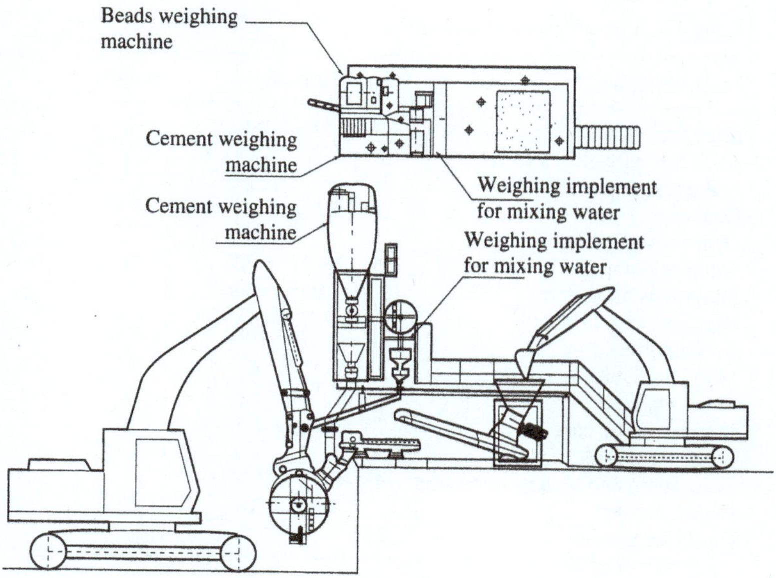

Figure 6.18. Beads treated soil application equipment.

Table 6.9. List of principal machines (EPS beads).

Machine name	Specifications		Number
Main unit	15 m^3/H		1
Back hoe	0.7 m^2		1
	0.4 m^2		2
	1.3 m^2		1
Cement silo	30 t		1
Generator	75 kVA	175 kVA	1
Distributing panel			1
Squeeze pump	NSP0030	70 m^3/h	1
Scale			1
High-pressure cleaner			1
Compressor			1
Sea water tank	30 m^2		1
Fresh water tank	3 m^2		1
Pouring crane	5 t		1
EPS beads tank			1
Blower			1

air foam treated soil and 16 days for the beads treated soil. The placing of the air foam and the beads treated soil are shown in Photographs 6.6 and 6.7 respectively.

After the completion of construction, the specimens were sampled to study the wet density and the unconfined compressive strength distributions. Figures 6.19 and 6.20 show the results.

Photograph 6.6. Placing of the air foam treated soil.

Photograph 6.7. Placing of the beads treated soil.

Figure 6.19 shows that six months after placing, the wet density of both the air foam treated soil and the beads treated soil had increased to about 0.05 g/cm^3 higher than the target wet density of 1.10 g/cm^3. It was assumed that this increase was due to the shrinkage of the treated soil during hardening and the increase of defoaming caused by the coagulant in the excavated shield soil used as the source soil. On the other hand, Figure 6.20 shows that at all depths, the unconfined compressive strength was far above the on-site minimum control value of 200 kN/m^2, and that the mean values were 610 kN/m^2 for the air foam treated soil and 540 kN/m^2 for the beads treated soil.

In using construction waste soil as the source soil in this case, due to the non-uniformity of its material properties, the mechanical characteristics of unconfined compressive strength showed scattered results, although the design values were satisfied.

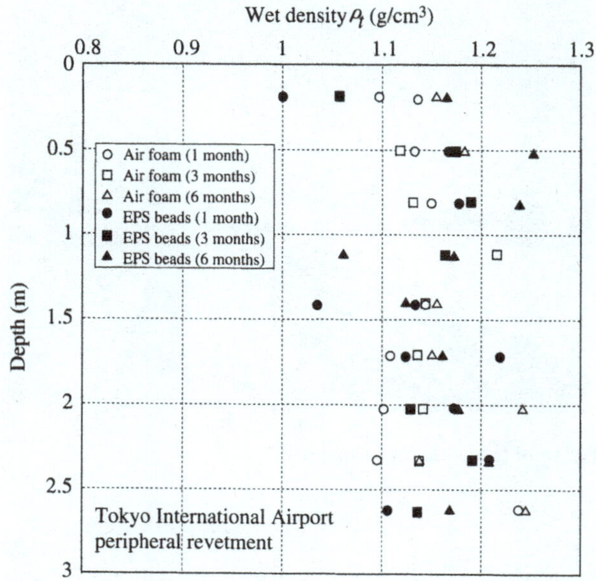

Figure 6.19. Depth distribution of wet density.

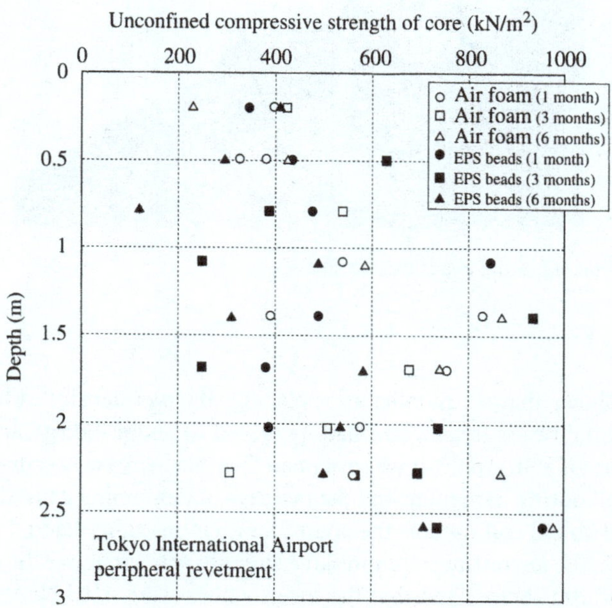

Figure 6.20. Depth distribution of unconfined compressive strength.

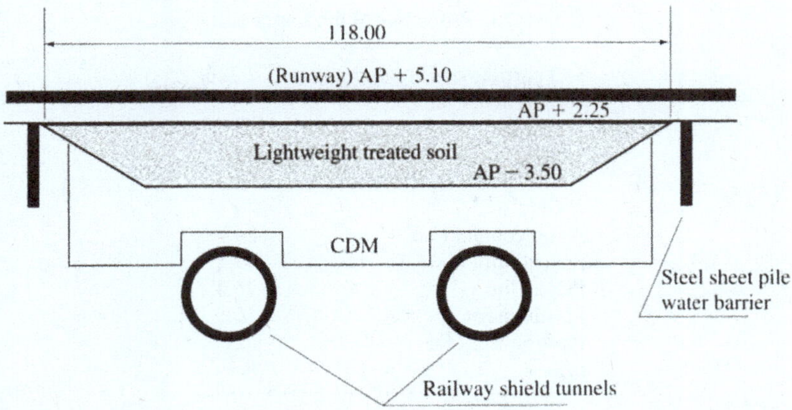

Figure 6.21. Cross-section of construction.

6.3.2 Case 2 [9][10]

It was thought that a layer susceptible to liquefaction existed above the tunnel at the intersection of the new parallel taxiway A and the railway tunnel extension to the airport terminal, that the embankment constructed for the taxiway would increase the load, and that the work would be affected because of its proximity to the railway shield tunnel work. So the ground above the shield tunnel was improved by the deep mixing soil stabilization method (CDM) and its upper part was replaced with air foam treated soil to reduce the overburden load on the tunnel and to prevent liquefaction. Figure 6.21 shows the cross-sectional view of its application.

Originally it was planned to adopt CDM to prevent liquefaction on the road bed of the taxiway, the embankment of the pavement, and the entire shield section. It was estimated that more than 50 cm of final settlement would occur due to the vertical pressure of the embankment and the CDM itself. Therefore, to balance the amount of settlement above the shield tunnel at that location with the quantity of settlement of the surrounding ground, CDM was applied down to a depth of AP −3.5 m and the soil above that level was replaced by lightweight treated soil. As a result of using the lightweight treated soil, it was possible to reduce the load increase by 60% compared to the case where the entire ground was treated by the CDM method. A design quantity of 32,162 m^3 of lightweight treated soil was used. The source soil used was construction soil with a high sand content. Table 6.10 shows the properties of the source soil.

The mix proportion was determined for a density of 1.2 g/cm^3 and design unconfined compressive strength of 200 kN/m^3. The foaming agent was a surface-active agent with 20× dilution and 25× foaming rates and the stabilizing agent was blast-furnace slag cement type B. Table 6.11 shows the mix proportion specifications.

Because the placing was carried out in a zone where the height is restricted due to passing aircraft, the use of cranes, etc. was prohibited. The air foam treated soil was pumped to the predetermined location by squeeze pumps, then fed from the pipe through a flexible pipe to a tremie pipe used to place it. Photograph 6.8 shows the placing, Figure 6.22 shows the application system, Figure 6.23 shows the plant facility layout, and Table 6.12 shows the principal machines that were used.

Table 6.10. Properties of the source soil.

Soil particle density (g/cm^3)	2.726
Soil content (%)	
Gravel	17
Sand	44
Silt	19
Clay	20
Water content (%)	24.9
Liquid limit (%)	43.2
Plastic limit (%)	26.4
Plastic index I_p	16.8
Ignition Loss (%)	7.3
pH	8.4

Table 6.11. Mix proportion specifications.

	Density (g/cm^3)	Dry soil (kg/m^3)	Water (kg/m^3)	Stabilizing agent (kg/m^3)	Lightweight material (l/m^3)
Air foam treated soil	1.1	507	456	120	322

Photograph 6.8. Placing in the field.

Specimens were collected after the construction to obtain their wet density and unconfined compressive strength. The results are shown in Figure 6.24.

The above results confirm that the wet density and the unconfined compressive strength both satisfy the design values in almost all the areas of placing. The average value for the unconfined compressive strength in particular was four times the design value. The strength of the material was high because the stabilizing agent content was increased slightly to increase the cohesiveness of the material in response to concern that the materials might

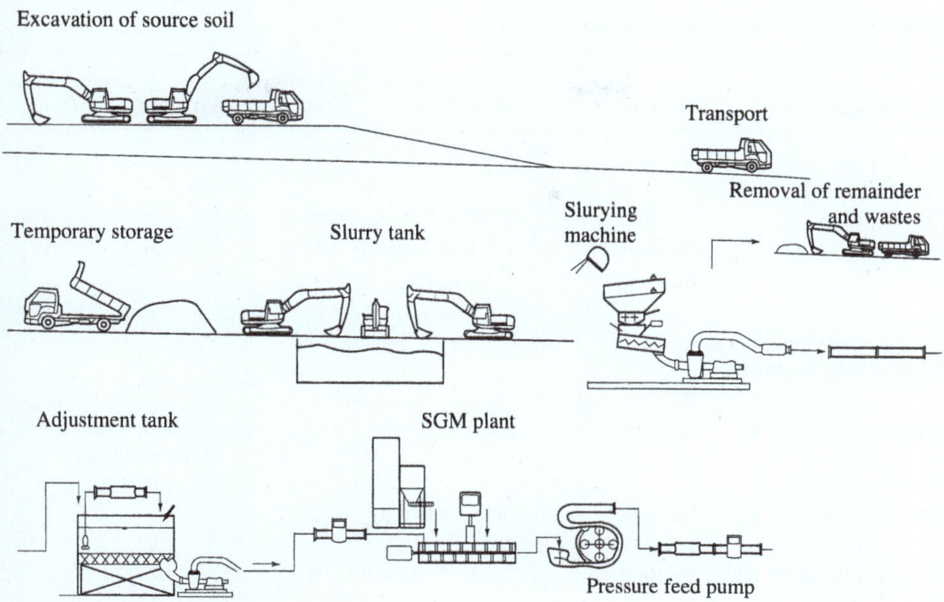

Figure 6.22. Application system.

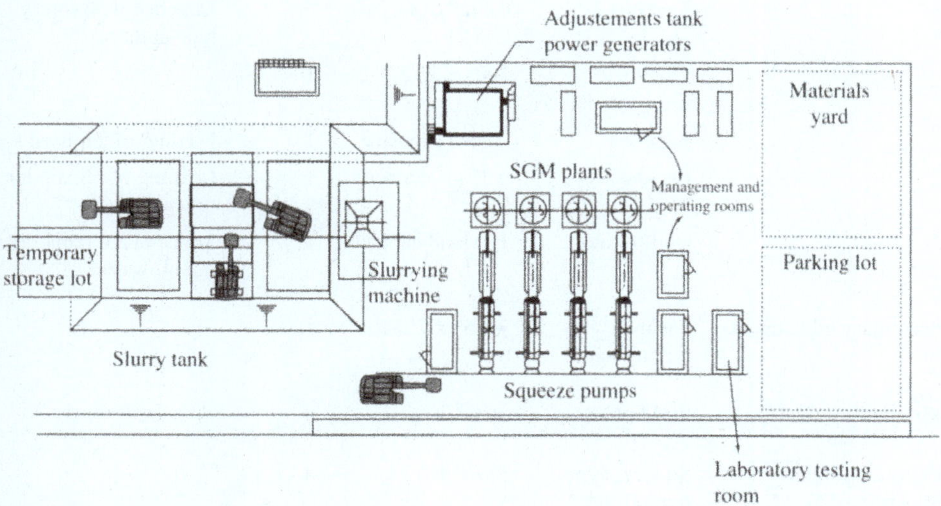

Figure 6.23. Plant layout map.

separate because construction soil with a high sand content was used as the source soil. The results of tests of the samples obtained the stipulated wet density. The results of measurements of vertical displacement of the shield tunnel after placing were satisfactory because the vertical displacement did not exceed the control value.

Lightweight treated soil was used for backfilling of the quaywall, but in this case, it was used to reduce the overburden load on the shield tunnel and to prevent liquefaction.

Table 6.12. List of principal machines.

Type of work	Machine	Specifications/ performance	Unit number	Remarks
Earth Work				
Excavation	Backhoe	0.6 m^3 class	4	
	Dump truck	10 t load capacity	6	
	Bulldozer	15 t capacity	1	
Soil Improvement Work (Lightweight treatment)				
Sorting of source soil	Backhoe	0.6 m^3 capacity	2	Attached with skeleton bucket
Loading of sorted soil	Backhoe	0.6 m^3 capacity	2	
Transport of sorted soil	Dump truck	10 t load capacity	4	
Primary pre-treatment	Tractor shovel	1.4 m^3 capacity	1	Pullout of sorted soil
	Backhoe	1.5 m^3 capacity	1	Mixing of adjusted soil
	Backhoe	0.6 m^3 capacity		Attached with rotary type agitator
	Vibratory slurrying device	200 m^3/h	1	
	Backhoe	2.0 m^3 capacity	1	Feeding of adjusted soil
	Backhoe	0.35 m^3 capacity	1	Loading of remainder or soil wastes
	Dump truck	10 t load capacity	1	Transport of remainder or soil wastes
Secondary adjustment	Secondary mixing tank	30 $m^3 \times 2$ tanks	1	
Mixing	SGM plant	50 m^3/h	4	
Pressure pumping transport of SGM	Squeeze type pump truck	55 m^3/h	4	
	Electric power generator	300 kVA	2	Slurrying device, secondary adjustment
	Electric power generator	220 kVA	4	SGM Plant
	Electric power generator	95 kVA	1	Controller
	Air compressor		3	Foaming device, valves, etc.

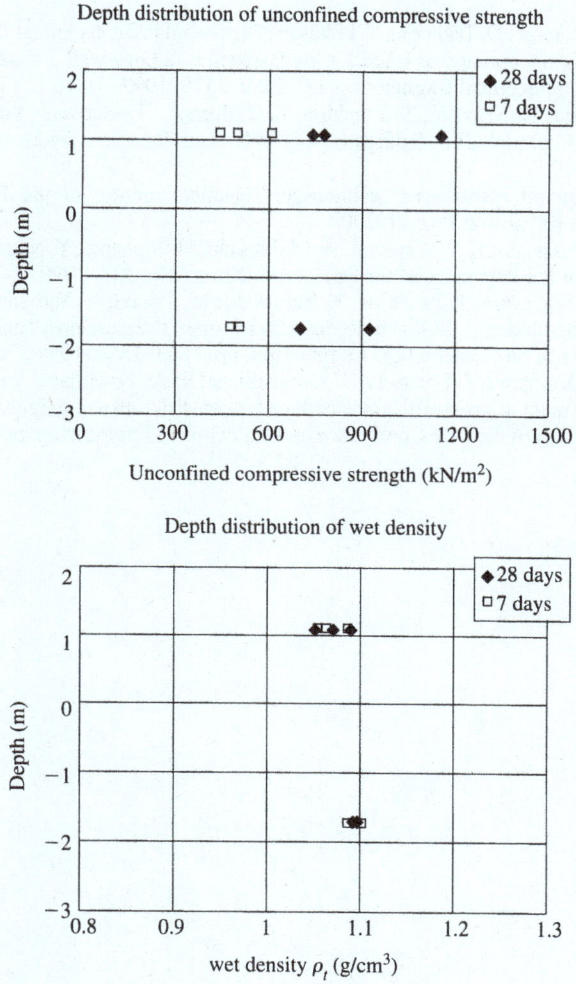

Figure 6.24. Depth distributions of the unconfined compressive strength and wet density.

REFERENCES

1. T. Tsuchida, K. Nagai, T. Okumura, T. Kishida, K. Funada: Mechanical Properties of Lightweight Geo-Material Used for the Backfill of a Quaywall (Part 1), Proceedings of the 31st Conference on Geotechnical Engineering, pp. 2525–2528, 1995.
2. T. Tsuchida, K. Nagai, M. Yukawa, T. Kishida, M. Yamamoto: Properties of Lightweight Soil Used for the Backfill of a Pier, Technical Note of the Port and Harbor Research Institute, No. 835, 1997.
3. T. Tsuchida: Development and Use of Foamed Treated Soil in Port and Airport Project, Report of the Port and Harbor Research Institute, Vol. 38, No. 2, pp. 131–167, 1999.
4. T. Wako, T. Tsuchida, Y. Matsunaga, K. Hamamoto, T. Kishida, T. Fukasawa: Use of Artificial Lightweight Materials (Treated Soil with Air Foam) for Port Facilities, *Civil Engineering Journal, No. 602/VI–40*, pp. 35–52, 1998.

5. T. Wako, Y. Matsunaga, D. Takeuchi, T. Fukasawa, T. Kishida: Geotechnical Properties of Foam-mixed Light-weight Geomaterial Used for the Backfill of a Quaywall, Proceedings of the 32nd Conference on Geotechnical Engineering, pp. 2569–2570, 1997.
6. S. Ifuku, K. Sakai, K. Miyoshi, T. Tsuchida, F. Hashimoto: Underwater Placing Test of SGM Lightweight Geomaterial, Proceedings of the 34th Conference on Geotechnical Engineering, pp. 761–762, 1999.
7. Coastal Development Institute of Technology: Technical records of the Tokyo International Airport Offshore Expansion Project, 2000.
8. T. Tsuchida, H. Fuzisaki, H. Nakamura, M. Makibuchi, H. Shinshia, Y. Nagasaka, Y. Hikosaka: Technical Note of the Port and Harbor Research Institute, No. 923, 1999.
9. H. Fuzisaki, K. Sugawara, T. Tsuchida, K. Yamasaki, M. Takazi, N. Yamane: New Application of Lightweight Geomaterial (SGM) to Reduce Overburden Pressure on a Tunnel, Proceedings of the 34th Conference on Geotechnical Engineering, pp. 1883–1884, 1999.
10. K. Sugawara, S. Kitagawa, T. Tsuchida, K. Yamazaki, S. Obara, N. Mihara: An example of estimation and measurement of ground displacement under construction using lightweight geomaterials (SGM), Proceedings of the 34th Conference on Geotechnical Engineering, pp. 1885–1886, 1999.